Synthetic Aperture Radar

J. Patrick Fitch

Synthetic Aperture Radar

C.S. Burrus, Consulting Editor

With 93 Illustrations

Springer-Verlag
New York Berlin Heidelberg
London Paris Tokyo

J. Patrick Fitch
Engineering Research Division
Electronics Engineering Department
Lawrence Livermore National Laboratory
University of California
Livermore, CA 94550
USA

Consulting Editor
Signal Processing and Digital Filtering

C.S. Burrus
Professor and Chairman
Department of Electrical and
  Computer Engineering
Rice University
Houston, TX 77251-1892
USA

Library of Congress Cataloging-in-Publication Data
Fitch, J. Patrick.
  Synthetic aperture radar.
  Includes bibliographies.
  1. Synthetic aperture radar.   I. Title.
TK6592.S95F58   1988   621.36′78   87-32110

Camera-ready copy prepared by the author using LaTeX.

9 8 7 6 5 4 3 2 1

ISBN-13:978-1-4612-8366-9   e-ISBN-13:978-1-4612-3822-5
DOI: 10.1007/978-1-4612-3822-5

PREFACE

Radar, like most well developed areas, has its own vocabulary. Words like Doppler frequency, pulse compression, mismatched filter, carrier frequency, in-phase, and quadrature have specific meaning to the radar engineer. In fact, the word radar is actually an acronym for RAdio Detection And Ranging. Even though these words are well defined, they can act as road blocks which keep people without a radar background from utilizing the large amount of data, literature, and expertise within the radar community. This is unfortunate because the use of digital radar processing techniques has made possible the analysis of radar signals on many general purpose digital computers. Of special interest are the surface mapping radars, such as the Seasat and the shuttle imaging radars, which utilize a technique known as synthetic aperture radar (SAR) to create high resolution images (pictures). This data appeals to cartographers, agronomists, oceanographers, and others who want to perform image enhancement, parameter estimation, pattern recognition, and other information extraction techniques on the radar imagery.

The first chapter presents the basics of radar processing: techniques for calculating range (distance) by measuring round trip propagation times for radar pulses. This is the same technique that sightseers use when calculating the width of a canyon by timing the round trip delay using echoes. In fact, the corresponding approach in radar is usually called the pulse echo technique. The second chapter contains an explanation of how to combine one dimensional radar returns into two dimensional images. A specific technique for creating radar imagery which is known as Synthetic Aperture Radar (SAR) is presented. Chapter 3 presents an optical interpretation and implementation of SAR. There are many similarities between SAR and other image reconstruction algorithms; a summary of tomography and ultrasound techniques is included as Chapter 4. Although the full details of these techniques are not explained, an intuitive understanding of the physical properties of these systems is possible from having studied the radar imaging problem.

Any type of digital radar processing will involve many techniques used in the signal processing community. Therefore a summary of the basic theorems of digital signal processing is given in Appendix A. The purpose of including this material is to introduce a consistent notation and to explain some of the simple tools used when processing radar data. Readers unfamiliar with the concepts of linear systems, circular convolution, and discrete Fourier transforms should skim this Appendix initially and refer to it as necessary. Matched filters are important in both pulse echo radar and SAR imaging: Appendices B and C discuss the statistical properties and digital implementation strategies for matched filters.

The approach in these notes is to present simple cases first, followed by the generalization. The objective is to get your feet wet, not to drown in vocabulary, mathematics, or notation. Usually an understanding of the geometry and physics of the problem will be more important than the mathematical details required to present the material. Standard techniques are derived or justified depending on which approach offers the most insight into the processing. Of course there are many radar related techniques which were simplified for presentation or omitted entirely—existing books and articles containing this information should be within the grasp of readers who studiously complete these notes.

These notes were initiated as part of the documentation for a software-based radar imaging system at Lawrence Livermore National Laboratory (LLNL). The code runs on a supercomputer developed in-house under the S-1 project. Some of the material presented here was also used in a graduate course at the University of California to introduce particular imaging systems and techniques. Comments by Lab researchers, faculty, and students have been helpful and encouraging during preparation of the manuscript. It is a pleasure to acknowledge my collaborators at LLNL: Steve Azevedo of the tomography research project and Jim Brase of the non-destructive signal processing program. Several of the figures in Chapter 4 were produced through joint efforts. Finally, a note of special appreciation and thanks to my wife Kathy for her encouragement and assistance with every aspect of the preparation of this manuscript.

Contents

Chapter 1

Radar Processing

1.1 Radar: A Well Defined Problem

Distance and time are equivalent. But don't panic, this is not going to be a discussion about Einstein's theory of relativity or the use of sundials. Actually, a simple example in everyday words is sufficient to describe how time measurements can be used to determine distances. Performing these types of measurements is the purpose of a radar system.

Suppose you are on a farm which has several open wells. Out of curiosity you might drop a pebble into one of the wells. After a few moments a splash or a dull thud is heard. The type of noise heard makes it possible to determine whether the well is wet or dry. The time from when the pebble was released until the noise was heard is proportional to the depth of the well. Obviously the well which has the longest drop to splash time is the deepest well. If the pebble dropping experiment is performed at each well, a plot of well depths is accumulated. The results for a six well farm can be seen in Figure 1.1. Note that the elapsed time, displayed on the horizontal axis, is sufficient for ranking the depth of any well relative to the other wells. Additional information would be required to calculate the absolute depth of any of the wells.

In a radar system the pebble is replaced by an electro-magnetic wave transmitted from an antenna. Another antenna, or in many cases the same antenna, serves as the "ear" waiting to hear a "splash". The splash corresponds to the reflection of the wave off some object in front of the antenna. By recording what the antenna receives it is possible to determine the relative position of objects which reflect the wave. A sample radar signal is given together with its return in Figure 1.2. For this example there are

1

Figure 1.1: Pebble dropping experiment.

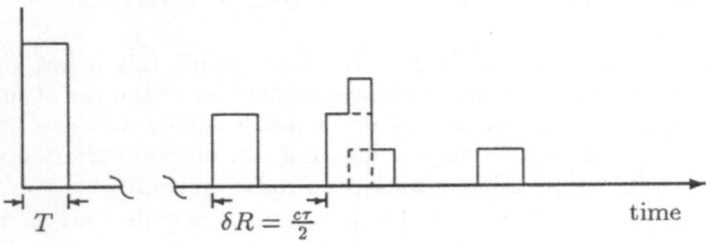

Figure 1.2: Sample radar reflection.

four reflecting objects or targets. Note that the reflections from two of the targets overlap to produce an irregularly shaped echo.

The vertical axis is a measure of how strongly the electro-magnetic wave was reflected off the objects and the horizontal axis displays the elapsed time from transmission of the pulse to reception. Some objects reflect radar signals very well and produce strong echoes—similar to the way a mirror reflects visible light. Because most objects do not reflect radar waves as efficiently as a mirror reflects light, only a portion of the incident radar energy is reflected. Because the electro-magnetic illumination is not from the visible region of the spectrum, the effect of the target on the radar wave (known as the target's signature) may not correspond with the human visual experience.

There are some important differences between the data obtained from the well experiment and the radar data. The pebble dropping experiment can be done one well at a time. In radar the number of wells is unknown and all the data is taken simultaneously. If two wells are found to have similar time delays, they would continue to be treated as separate while evaluating whether they are wet or dry. The format of the graphical display of Figure 1.1 may need to be changed, but all the information about every well is still obtained. Suppose the wells are close enough to permit simultaneously dropping a pebble in every well. Now an experiment which is similar to radar can be performed with a microphone and tape machine recording the splashes and thuds. This approach yields the same results as the sequential stone dropping experiment unless some of the splashes or thuds overlap during recording. Additional information would be required to distinguish between two simultaneous small splashes and one big splash. The problem becomes even more difficult when the possibility of a splash covering up a thud is considered. A radar antenna illuminating two objects with approximately the same time delay will create a similar problem. The reflections off the two objects may add and create a return signal that appears as one object at that time delay with a larger reflectivity. This problem is compounded by the random fluctuations in radar signals due to interfering radiation, atmospheric effects, thermal changes in the electronic components and other unpredictable degradations of the signal. The processes which contribute to the degradation from ideal transmission and reception of the radar signals are called noise. Under anything but ideal (noiseless) conditions it is impossible to determine the precise number or nature of objects in the radar return. The corresponding signal is therefore considered ambiguous—having at least two possible interpretations.

The reason ambiguous signals are received is the overlap of the returning radar pulse from closely spaced objects. Clearly, making the pulses shorter in duration, will reduce the ambiguity caused by overlapping reflections. However, as long as the pulses have some width there will be some minimum time delay between targets which is necessary to have unambiguous reception. In fact, to guarantee non-overlapping reflections, targets must be separated in time delay by at least the width of the transmitted pulse.

If the radar's transmitted pulse duration is T seconds, then the time delay between objects must be at least T seconds to prevent interference between the pulse echoes. Because radar waves are a form of electromagnetic radiation, they travel at a constant velocity c equal to the speed of light. This constant has been measured experimentally as approximately $3 \times 10^8 m/sec$. If the time delay between the reflection off two objects is τ, the light (radar pulse), had to travel an additional $c\tau$ meters for the farther object. Because $c\tau$ represents a round trip distance for the pulse, the actual physical separation of the objects is $c\tau/2$. For objects separated by less than $cT/2$, the reflections will overlap making it difficult to resolve where one target ends and the other begins. For this reason, $cT/2$ is usually called the range resolution of the radar system. These parameters were defined graphically in Figure 1.2.

Reducing the duration T of the radar pulse, improves the system's range resolution $cT/2$, which results in the radar being capable of discerning objects which are closer together. In the well experiment, the reduction in pulse width might correspond to using smaller pebbles which result in smaller (quieter and shorter duration) splashes. This creates a new problem: if the splash becomes too short, it will not create enough sound for the microphone. For radar signals, the loudness of the splash corresponds to the energy in the reflected pulse. The farther the targets are from the antenna the more energy is required in the pulse. Unfortunately it is more difficult and consequently more expensive to build a radar transmitter and receiver for a short pulse than for a long pulse of equal energy. What is needed is a transmitted pulse of sufficient duration to maintain the required energy levels together with a clever means of receiving and processing this pulse so that the data can be treated as if it were from a short pulse. In simpler terms: design a pulse so that overlapping returns from different time delays can be separated.

Just as the shape and size of the pebbles being dropped in the well can be controlled, the shape and energy in the transmitted radar light wave can be controlled. Let $u(t)$ be the function which describes the shape of the pulse. If the pulse is of length T, then $u(t) = 0$ for $t < 0$ and $t > T$. An object at time delay τ_1 will reflect a signal of the form $r_1(t) = \sigma_1 u(t - \tau_1)$, where σ_1 is some real-valued[1] number representing how strongly the object reflects the transmitted light. A second target might have a reflection of the form $r_2(t) = \sigma_2 u(t - \tau_2)$. Recall that the σ's correspond to the type of noise the pebble creates (splash or thud) and the τ's correspond to the elapsed time from drop to splash. The goal is to design a pulse shape $u(t)$ such that the returns are dissimilar for objects at different distances from the radar antenna—that is, when $\tau_1 \neq \tau_2$.

One possible measure of the similarity of the two waveforms r_1 and r_2 is the squared difference d^2 defined as

$$d^2 = \int \left[r_1(t) - r_2(t) \right]^2 dt. \tag{1.1}$$

Similar signals will have a relatively small squared difference. For example, the squared difference of a signal with itself is zero. Note that this equation defines what is meant by similar and dissimilar. There are many other

---

[1] In order to completely characterize the reflectivity σ of a target, a complex-valued representation with dependencies on illumination orientation and polarization is necessary. For simplicity, the orientation dependencies of σ are not considered in our analysis.

intuitively satisfying definitions for measuring similarity. For instance, the squared operator in the integral of Equation 1.1 could be replaced with an absolute value operator raised to an arbitrary power greater than one. For the mathematics which follows, however, the squared difference is perhaps the most convenient definition. Continuing with our design, we want to maximize the squared difference everywhere τ_1 does not equal τ_2. Expanding d^2 results in

$$
\begin{aligned}
d^2 &= \int [r_1(t) - r_2(t)]^2 dt \\
&= \int [\sigma_1 u(t - \tau_1) - \sigma_2 u(t - \tau_2)]^2 dt \\
&= \sigma_1^2 \int [u(t - \tau_1)]^2 dt + \sigma_2^2 \int [u(t - \tau_2)]^2 dt \\
&\quad - 2\sigma_1 \sigma_2 \int u(t - \tau_1) u(t - \tau_2) dt.
\end{aligned}
\tag{1.2}
$$

The first two terms which represent the energy in the reflections depend on $u(t)$ only through its energy $\int [u(t)]^2 dt$. Because the energy of the pulse can be controlled by scaling the pulse shape $u(t)$, these energy terms can be omitted from the optimization. The target reflectances, σ_1 and σ_2, are functions of the objects observed and cannot be predicted for designing the pulse shape. The maximization of d^2 has been reduced to minimizing the integral

$$
A(\tau_1, \tau_2) = \int u(t - \tau_1) u(t - \tau_2) dt
\tag{1.3}
$$

for $\tau_1 \neq \tau_2$. For $\tau_1 = \tau_2$, the reflections overlap exactly and represent two targets at an identical time delay from the antenna. For now we will assume that these types of targets cannot be discriminated and their squared difference should be large to denote reflection from objects at the same time delay (distance from the antenna). The transmitted pulse shape $u(t)$ should therefore be selected to maximize $A(\tau_1, \tau_2)$ for $\tau_1 = \tau_2$ and should fall off quickly for $\tau_1 \neq \tau_2$. This function $A(\tau_1, \tau_2)$, which is called the autocorrelation function, equals the energy in $u(t)$ for τ_1 equal τ_2.

This analysis shows that in order to have good resolution, the autocorrelation should be small everywhere except when the time delay between signals is zero. Note that by a change of variables the autocorrelation can be written

$$
A(\tau) = \int u(t) u(t + \tau) dt
\tag{1.4}
$$

where τ can equal $\tau_1 - \tau_2$ or $\tau_2 - \tau_1$. The limits on the integral would be over one cycle for periodic signals, all allowed values for finite length signals, and the limit of a symmetric average for signals of infinite extent. Integration determines the area under the product of $u(t)$ with $u(t + \tau)$, where $u(t+\tau)$ is a shifted version of $u(t)$. $A(\tau)$, therefore, is the value of the integral for a particular shift τ. The autocorrelation for several functions is shown in Figure 1.3. Rectangular pulses are especially helpful when developing an intuition for the shift and integrate function performed by the autocorrelation operation.

The correlation operation, which is similar to the autocorrelation, is used when two different functions are to be compared. For two possibly complex functions u and v, the correlation is defined mathematically as

$$C(\tau) = \int u^*(t)v(t + \tau)dt \qquad (1.5)$$

where $*$ is the complex conjugate operation. In order to simplify the mathematics, it is often convenient to represent radar signals as complex functions. The correlation of two functions can be considered as a plot of their average product as one function is slid past the other. An autocorrelation occurs when one function is used for both inputs to a correlation. Note that if the waveforms r_1 and r_2 in Equation 1.2 had not been scaled versions of the pulse shape u, then the maximization would have been reduced to minimizing the correlation of r_1 with r_2. When two signals which are known to be different are compared using a correlation operation, the output C is often referred to as the cross-correlation. The term correlation refers generically to the comparison of two arbitrary signals by integrating the product of the two signals for different relative shifts.

Functions of the form $\cos[2\pi(ft + .5at^2)]$ or more generally $e^{j2\pi(ft + .5at^2)}$ compress into very sharp autocorrelations. As an example, consider the autocorrelation of the complex exponential with phase $2\pi(ft + .5at^2)$. The first time derivative of this phase is $2\pi(f + at)$, which is the equation for a line of slope a and initial value f. This first derivative of the phase is called the frequency of the waveform. The 0.5 coefficient of a in the phase is used so that the frequency will be a line of slope a. This type of signal is therefore called a linear frequency modulated (linear fm) signal. The larger the value of a the faster the frequencies change. Because it is a burst of different frequencies, the linear fm signal is also called a chirp of rate a. The autocorrelation of a linear fm signal can be calculated as

$$
\begin{aligned}
A(\tau) &= \int e^{-j2\pi(ft + .5at^2)} e^{j2\pi(f(t+\tau) + .5a(t+\tau)^2)} dt \qquad (1.6) \\
&= e^{j2\pi(f\tau + .5a\tau^2)} \int e^{j2\pi a\tau t} dt.
\end{aligned}
$$

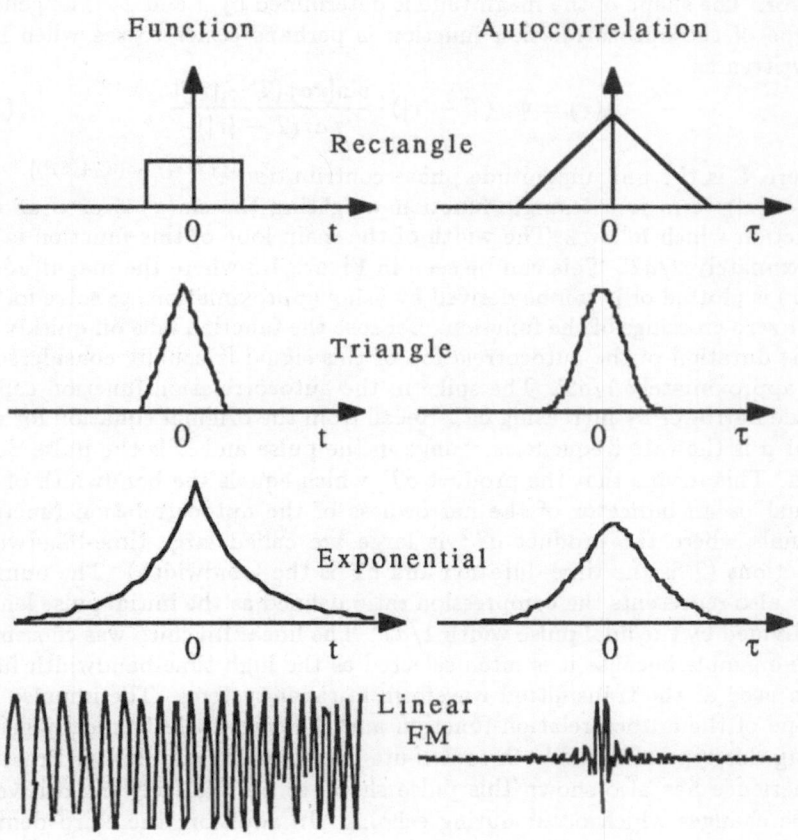

Figure 1.3: Autocorrelation of several functions.

In order to evaluate the integral, let the complex exponential be a pulse of duration T beginning at time T_0. This results in

$$A(\tau) = e^{j2\pi\tau[f+a(T_0+.5T)]} \frac{\sin[\pi a\tau(T-|\tau|)]}{\pi a\tau} \text{ for } -T \le \tau \le T. \quad (1.7)$$

The initial frequency $f + aT_0$ contributes to the autocorrelation as a phase factor. The shape of the magnitude is determined by a and T. The general shape of the autocorrelation function is perhaps easier to see when it is rewritten as

$$A(\tau) = \Phi \cdot (T - |\tau|) \cdot \frac{\sin[\pi a\tau(T-|\tau|)]}{\pi a\tau(T-|\tau|)} \quad (1.8)$$

where Φ is the unit magnitude phase contribution $e^{j2\pi\tau[f+a(T_0+.5T)]}$. The $(T - |\tau|)$ term is a triangle function weighting the $\sin(x)$ over x or sinc function which follows. The width of the main lobe of this function is approximately $2/aT$. This can be seen in Figure 1.3 where the magnitude of $A(\tau)$ is plotted or it can be derived by using approximations to solve for the first zero crossings of the function. Because the function falls off quickly the time duration of the autocorrelation of this signal is usually considered to be approximately $1/aT$. The spike in the autocorrelation function can be made narrower by increasing aT. Recall from the original equation for $u(t)$ that a is the rate frequencies change in the pulse and T is the pulse duration. This means that the product aT, which equals the bandwidth of the signal, is an indicator of the narrowness of the autocorrelation function. Signals where the product aT^2 is large are called large time-bandwidth functions (T is the time-duration and aT is the bandwidth). The number aT^2 also represents the compression ratio defined as the initial pulse length T divided by the final pulse width $1/aT$. The linear fm chirp was chosen for this example because it is often selected as the high time-bandwidth function used as the transmitted waveform in radar systems. The impulse like shape of the autocorrelation function and the ease of electronically generating this type of signal both contribute heavily to its popularity. Practical experience has also shown this pulse shape to be relatively insensitive to scale changes which occur during echoing. In addition, the chirp demonstrates the accepted radar axiom that the system range resolution can be improved by increasing the bandwidth of the transmitted pulse.

Selection of a pulse shape with an autocorrelation function which is large near zero and falls rapidly implies that returns which partially overlap can be separated as reflections from different objects. In fact, the autocorrelation function not only defines the measure of how similar partially overlapping returns will be, but also provides a method for separating these reflections. For a pulse shape $u(t)$ and received signal of the form

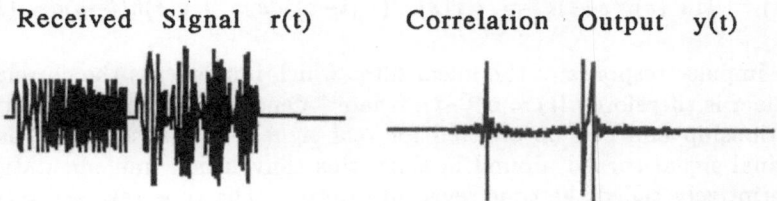

Figure 1.4: Detection of pulses using the autocorrelation.

$r(t) = \sigma u(t - \tau)$, the receiver that implements a correlation for complex signals is given by

$$y(t) = \int u^*(s)\, r(s+t)ds = \sigma \int u^*(s)\, u(s+t-\tau)ds = \sigma A(t - \tau) \quad (1.9)$$

The output $y(t)$ of the receiver will be large when t equals τ and will be small otherwise—assuming the pulse shape $u(t)$ was selected using the correlation criteria. This means that the output of this receiver will have spikes associated with time delays which correspond to reflecting objects. The transmitted pulse waveform $u(t)$ has been compressed to a spike which has the shape of its autocorrelation function $A(\tau)$. An example of a reflected signal with overlapping returns from targets with equal reflectivity is given in Figure 1.4. Because the correlation receiver performs a linear operation, pulses reflected with more energy will result in larger spikes after reception than spikes resulting from lower energy reflections.

In general, if there are N targets reflecting energy, then there will be N spikes output from the correlation receiver with each one scaled based on the reflectivity of the associated target. To summarize, if $r(t)$ is given by $\sum_{i=1}^{N} \sigma_i u(t - \tau_i)$, then the output of the correlation receiver is

$$\begin{aligned} y(t) &= \int u^*(s)\, r(s+t)\, ds \\ &= \sum_{i=1}^{N} \int u^*(s)\, u(s+t-\tau_i)\, ds \\ &= \sum_{i=1}^{N} \sigma_i\, A(t - \tau_i) \end{aligned}$$

The linearity of the correlation operation implies that this receiver can be expressed as the convolution of the received signal with an impulse response $h(t)$. Expanding the correlation receiver implementation

$$y(t) = \int u^*(s)r(s+t)ds = \int r(s)u^*(-(t-s))ds = \int r(s)h(t-s)ds. \quad (1.10)$$

The impulse response of the linear filter which implements the correlation receiver is therefore $h(t) = u^*(-t)$, where * denotes the complex conjugate relationship and can be omitted for real signals. Because $u^*(-t)$ is the original signal turned around in time, this convolution implementation is descriptively called the time reversed receiver. The time reversed receiver can be easily implemented in an existing linear filter architecture by modifying the impulse response. Of course frequency domain implementations of linear filters are often the most efficient and should be investigated. The Fourier transform of the receiver output $y(t)$ can be calculated as the product of $R(f)$ and $H(f)$ which are the Fourier transforms of the received signal and the impulse response, respectively. The frequency response of the system can be expressed in terms of the Fourier transform of the signal $u(t)$ which is denoted by $U(f)$. This relationship is

$$
\begin{aligned}
H(f) &= \int h(t)\,e^{-j2\pi ft}\,dt \\
&= \int u^*(-t)\,e^{-j2\pi ft}\,dt \\
&= \left\{ \int u(t)\,e^{-j2\pi ft}\,dt \right\}^* \\
&= U^*(f).
\end{aligned}
$$

The result is that this filter can be implemented in the frequency domain as

$$Y(f) = R(f)\,H(f) = R(f)\,U^*(f). \quad (1.11)$$

This is known as the conjugate receiver implementation. Note that even for real signals $u(t)$, the Fourier transform $U(f)$ will be complex and require the complex conjugate when implementing this receiver. Another interpretation is that the complex conjugate operator in the frequency domain corresponds to a complex conjugate together with time reversal in the time domain. The autocorrelation operation for a signal $u(t)$ is equivalent to taking the magnitude squared $|U(f)|^2$ of the signal's Fourier transform.

The filter described by the correlation, time reversed, and conjugate receivers is referred to as the matched filter. This is because the filter is essentially a reference replica of the transmitted pulse which is compared to the received signal. When characteristics of the propagation medium are known, the reference waveform is given the shape of the anticipated received signal. The filter output is a measure of how precisely the received signal and the reference match. It can be proven that the matched filter is statistically the optimum filter for performing this operation under certain conditions. Because the proof is not necessary for understanding the implementation of matched filters, it is relegated to the Appendices.

We now have the most basic and powerful capability for our radar system: the ability to transmit a pulse of sufficient energy to withstand the round trip attenuation without sacrificing range resolution. The signals used so far have been represented as continuous functions of time. Because the concern here is with the implementation of a radar system on a digital computer, the continuous time signals must be sampled before beginning processing. An explanation of the digital signal processing of radar signals, however, requires first describing several additional details of analog (continuous time) radar systems.

1.2   Transmitting and Receiving

Once a pulse shape $u(t)$ has been selected, the next step is to transmit this pulse using an antenna. Typically, the pulse is translated up to a higher frequency, known as the carrier frequency, before transmission. This is exactly what radio and television stations do when transmitting their signals. If these signals were not transmitted at different carrier frequencies, they would interfere with each other. Occasionally this interference can be heard when two AM radio stations are received at the same frequency. Because different carrier frequencies possess different advantages and disadvantages, electro-magnetic waves are classified according to their frequencies. For instance, low frequencies (LF) are often used for navigational radio beams, very high frequencies (VHF) are used for television, police communications, and FM radio. The intermediate frequencies between LF and VHF are used in AM radio and many other communication systems. The range, fidelity, and interference conditions determine which frequencies are best for which applications. Radar systems usually use carriers that are higher frequency than VHF signals. This region of the spectrum is commonly called microwaves. Microwave systems are also used for satellite and telephone links. Microwaves are well suited for transmission over long distances without significant degradation due to atmospheric interference. This means that microwave radar systems should operate satisfactorily even during periods of precipitation or with heavy cloud cover. It should be noted that radar systems have been implemented at many different frequencies and each system has frequency dependent advantages and limitations. The microwave frequencies have provided reliable performance for radar systems.

In order to take advantage of the transmission properties of different frequencies, the pulse $u(t)$ is translated to a new frequency. Call this new frequency the carrier frequency f_c. As was established in an earlier section, the chirp $u(t) = \cos[2\pi(ft + .5at^2)]$ is a useful pulse shape for radar signals. Consequently, $s(t)$ could be generated directly as $\cos[2\pi(f_ct + ft + .5at^2)]$. Usually sinusoidal signals are translated from one carrier frequency to another using a procedure known as mixing. To do this, $u(t)$ is multiplied by a cosine or sine of frequency f_c and then filtered to pass only the desired frequency terms. Several trigonometric identities will be useful during this discussion. Recall that

$$
\begin{aligned}
2\cos(\alpha)\cos(\beta) &= \cos(\alpha - \beta) + \cos(\alpha + \beta) \\
2\sin(\alpha)\cos(\beta) &= \sin(\alpha - \beta) + \sin(\alpha + \beta) \\
2\sin(\alpha)\sin(\beta) &= \cos(\alpha - \beta) - \cos(\alpha + \beta)
\end{aligned}
\tag{1.12}
$$

These equations show that if the initial or base band pulse function is a sinusoid, then multiplication with another sinusoid results in a pair of sinusoids with sum and difference frequencies. By selecting the appropriate frequency term the original signal can be translated to a new frequency. For a linear fm signal $u(t)$, multiplication with $2\cos(2\pi f_c t)$ results in

$$
2\cos(2\pi f_c t)u(t) = \cos[2\pi(f_c t - ft - .5at^2)] + \cos[2\pi(f_c t + ft + .5at^2)]. \tag{1.13}
$$

The second term represents the sum of the frequencies and the first term represents the lower frequency difference. This assumes the frequency values f_c and f are positive. The first term can be eliminated by passing the signal $2u(t)\cos(2\pi f_c t)$ through a high pass filter. The term high pass means that higher frequency components of the signal are allowed to pass through the filter while lower frequency components are attenuated as much as possible. This results in

$$
s(t) = \cos[2\pi(f_c t + ft + .5at^2)] \tag{1.14}
$$

being transmitted. The frequency (time derivative of the phase) of $s(t)$ is $2\pi(f_c + f + at)$. This implies that the frequency of $u(t)$ has been increased by f_c or $2\pi f_c$ if radial frequency is used. The process of multiplying by a sinusoid and then filtering to select the desired frequency term is known as mixing. Successful mixing requires the signal to have a bandwidth (width in the frequency domain) which is small relative to the carrier frequency used in the mixing. If this condition is not met, the filtering operation will attenuate portions of the desired signal in the frequency domain. As was previously mentioned, the signal $s(t)$ may be created by mixing different frequencies or it may be generated directly.

Now consider the reflection of the transmitted signal off an object at distance R from the antenna. This means the signal pulse traveled a distance $2R$ from transmission to reception. Using the speed of propagation c, this distance corresponds to a time delay of $\tau = 2R/c$. Therefore, the received signal is of the form $s(t - \tau)$. Because the carrier frequency f_c was introduced to improve transmission and reception of the wave, not to change the functional form of the pulse, it may be desirable to remove the effect of the carrier frequency from the received signal. This translation in frequency, again, can be done by mixing. The received signal $r(t) = s(t - \tau)$ is multiplied by a sinusoid and then filtered to obtain the desired frequency components. This is often called beating the signal down to base band. It may be convenient to translate the received signal down to a lower frequency but not all the way to base band. This intermediate frequency is called the IF and will be denoted by f_{if}. Note that the translation to base band is obtained by setting f_{if} to zero. To see this, examine the multiplication of the received signal with a cosine of frequency $f_c - f_{if}$ to generate the beaten signal r_b:

$$
\begin{aligned}
r_b(t) &= 2r(t)\cos[2\pi(f_c - f_{if})t] \\
&= 2s(t - \tau)\cos[2\pi(f_c - f_{if})t] \qquad\qquad (1.15) \\
&= \cos[2\pi(-f_c\tau + f_{if}t + f(t - \tau) + .5a(t - \tau)^2)] \\
&\quad + \cos[2\pi(-f_c\tau + (2f_c - f_{if})t + f(t - \tau) + .5a(t - \tau)^2)].
\end{aligned}
$$

In order to translate the beaten signal to a lower frequency, the higher frequency term is attenuated (low pass filtered). The filtering operation, shown in Figure 1.5, produces a signal

$$
r_{if}(t) = \cos[2\pi(-f_c\tau + f_{if}t + f(t - \tau) + .5a(t - \tau)^2)]. \qquad (1.16)
$$

which equals the lower frequency term in the previous equation. Note that the different frequency terms will overlap if the bandwidth of the transmitted pulse is greater than $2(f_c - f_{if})$. This condition, referred to as the narrow band assumption, is required if a linear low pass filter is used to separate the two signal components.

The sinusoids for mixing operations are produced using a STable Local Oscillator or STALO (pronounced *stay-low*). When the same STALO is used to produce the mixing sinusoids for both transmission and reception, the phase as well as the amplitude and frequency of the received signal contains information about the targets. Systems which permit the extraction of phase information are called coherent systems. The $-f_c\tau$ term in the argument of the cosine in Equation 1.16 is the coherent phase information produced in this example. Essentially, the intermediate frequency

a) Received   Mixed   Output at IF
   Signal r(t)   Signal   Frequency

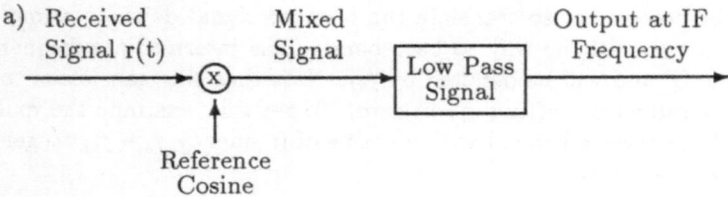

                    Reference
                    Cosine

b) Received Signal   Mixed Signal   Output Signal

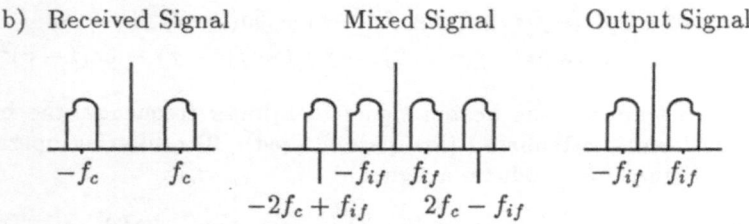

Figure 1.5: Coherent reception by mixing sinusoids: a) schematic representation and b) frequency domain operation.

is the new carrier frequency which could have been adjusted to zero (or any value) at reception. If the time scale had not been maintained by the STALO, the variable t would not represent the same number in both the transmitter and receiver equations. This means the $f_c t$ terms would not have cancelled exactly and $-f_c \tau$ would not be recoverable. In simple terms, the time origin must be maintained for both transmission and reception in order to have a coherent radar system. The time (or length if range units are used) for which the signals and the reference remain stable enough to allow extraction of phase information is called the coherence time (length). Readers with an optics background may notice some similarity between coherent reception for maintaining phase information and holographic techniques. Optical interpretations are discussed in more detail in Chapter 3.

Often two channels are run simultaneously—one channel performing multiplication by a cosine and the other channel multiplying with the same cosine shifted in phase. By shifting the phase of the second cosine one quadrant (90 degrees or $\frac{\pi}{2}$ radians), the zeros of the first cosine align with the peaks of the shifted cosine. This allows the information in the received signal, which would have been lost due to the zeros in the reference cosine, to be preserved in the other channel. Recalling that $\cos(\theta - \frac{\pi}{2})$ equals $\sin(\theta)$, we see that this means one channel is multiplied by a cosine and the other channel by a sine. Using Euler's relationship of $e^{j\theta} = \cos(\theta) + j\sin(\theta)$, these two multiplications can be thought of as one multiplication converting from real to complex numbers. Because the cosine term is assigned to the real axis, it is commonly called the in-phase term. The second channel, corresponding to multiplication by the sine, aligns with the imaginary axis. Because this is a one quadrant shift from the cosine term, the sine channel is called the quadrature term. Each complex number represents the two channel system with the real axis being the in-phase and the imaginary axis the quadrature. In order to keep only the lower frequency terms after reference multiplication, the two channels are each low pass filtered. The conversion from a one channel real signal to a complex representation can also be done later during the digital processing using Hilbert or Fourier transform techniques. That is, the conversion to in-phase and quadrature (I-Q) format does not need to be done during the analog mixing. This is an important convenience which reduces the required analog circuitry while retaining the ability to convert to a complex number representation for the phase extraction provided by the two channel coherent system. The various mixing procedures for a radar system are summarized in Figure 1.6.

Note that the reflected signal is actually a scaled version of the transmitted signal. That is, $r(t) = \sigma s(t - \tau)$. The factor σ is a function of the round trip propagation distance (which depends on τ), as well as the carrier

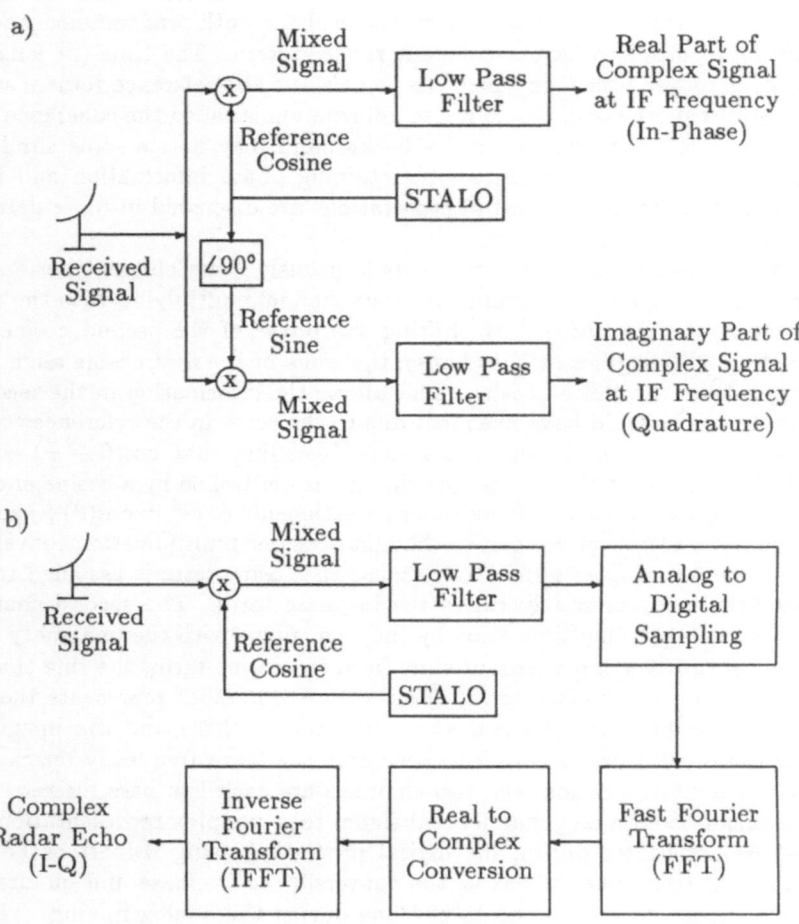

Figure 1.6: Different receiver systems: a) Two channel analog and b) Digital conversion of one channel analog signal to two channels.

frequency f_c, and the propagation medium (usually air). Combining these effects into one multiplicative coefficient is a reliable approximation which is usually made in radar. The reflection coefficient σ can be thought of as the "color" of the object as recorded by the radar system while the object was illuminated with light of frequency f_c. When this type of return signal is processed by a matched filter, the compressed output will be scaled by some σ and located at a time delay proportional to τ. These are both results of the linearity of the matched filter and are expressed in Equation 1.9. The final processed radar return will therefore have an amplitude which is proportional to the reflectivity of the targets in front of the antenna. The time delay of this processed signal is proportional to the round trip distance between antenna and targets. The shape of each compressed pulse will be the autocorrelation of the transmitted waveform—recall the correlation receiver implementation. The autocorrelation for linear fm signals was derived earlier in Equation 1.7. The shape of this autocorrelation (a narrow main lobe with several neighboring lobes of lower amplitude known as side lobes), often requires additional processing to reduce the size of the side lobes. Techniques known as windowing can be used to reduce side lobe amplitude at the expense of broadening the main lobe. The windowing operation is a simple multiplication of the frequency coefficients by the window weights—an example of frequency domain windowing for the autocorrelation of a linear fm waveform and a set of sinusoidal window weights is shown in Figure 1.7. Even when no windowing has been utilized explicitly, the data is always weighted by a rectangular window with constant value one. Ringing (side lobes) are reduced by tapering the edges of the rectangular window, this decreases the sharpness in the frequency domain which broadens the impulsive main lobe in the time domain. Windowing is analogous to slowly tapering off to zero the coefficients of a Fourier series rather than using an abrupt truncation which results in Gibb's phenomena.

The analog or continuous time processing of radar signals can be segmented into the steps of pulse generation, mixing up to a carrier frequency f_c, transmission, reflection, reception, mixing down to an intermediate frequency f_{if}, and pulse compression with a matched filter. For cases when f_{if} is nonzero, the pulse shape used as the reference in the matched filter receiver must be shifted up to f_{if}. In many digital radar systems the processing begins with the implementation of the matched filter on one channel of a received signal mixed down to some intermediate frequency (IF). Clearly if the signal is beat down to base band, the same approaches can be used with f_{if} equal to zero. If the sampling is performed at a later stage of the analog system, then the corresponding stages of the digital system can be omitted.

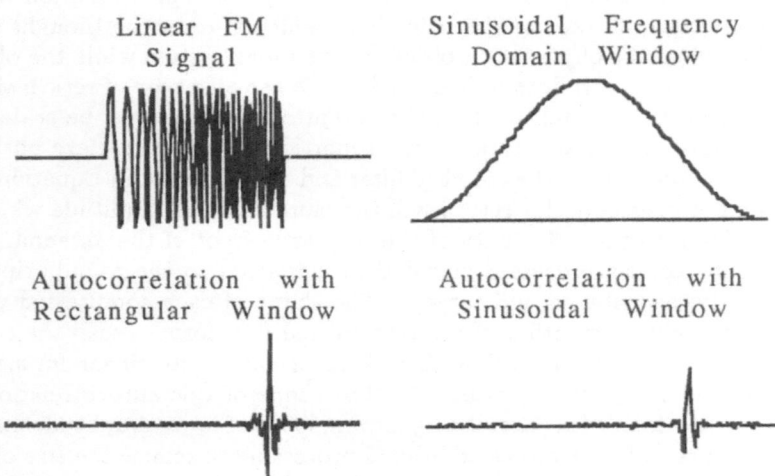

Figure 1.7: Frequency domain windowing to reduce side lobes.

1.3   Digital Processing of Radar Returns

The digital processing of the radar signal begins with the sampling of the received analog data. Assume that the sampling is performed on one channel of radar data which has been mixed down to an intermediate frequency f_{if}. This is a general case which includes mixing down to base band ($f_{if} = 0$) and no mixing at all ($f_{if} = f_c$). When two coherent analog channels (inphase and quadrature) are used to receive the signal, the corresponding digital operation which converts from real to complex representation can be omitted. Many of the analog and digital operations can be interchanged to adapt to the hardware particular to a given radar system.

The signal representing the mixed down received reflection of a linear fm chirp of rate a, initial frequency f, duration T, carrier frequency f_c, intermediate frequency f_{if}, and reflectivity σ is given by

$$r_{if}(t) = \sigma \cos[2\pi(-f_c\tau + f_{if}t + f(t-\tau) + .5a(t-\tau)^2)]\Re_T(t-\tau). \quad (1.17)$$

The rectangle function $\Re_T(t)$ is defined as one for t in the interval $[0, T]$ and zero outside of this interval. The appropriate reference pulse for the matched filter is therefore given by

$$u(t) = \cos[2\pi(f_{if}t + ft + .5at^2)]\Re_T(t). \quad (1.18)$$

Note that the mixing has shifted the frequency of the original pulse waveform by f_{if}. These two analog signals r_{if} and u are the basis for the digital implementations which follow.

Suppose that sampling begins at some time T_i and ends at time T_f. These are measured assuming the time origin to be when the pulse transmission begins. The constant propagation speed c of the radar pulse implies that the sampling interval corresponds to reflections from objects located at ranges between $cT_i/2$ and $c(T_f - T)/2$. The division by two in both terms accounts for the round trip propagation. The $-T$ is required in the second term so that an entire pulse waveform will be contained in the received signal. Segments of reflected pulses will result in imprecise matched filtering which cannot be regarded as valid data. This discussion has proceeded in an order reversed from a typical design of radar system specifications. In the the typical radar design an initial range of interest R_i and a final range of interest R_e are specified first. These distances imply sampling in time from $T_i = 2R_i/2$ to $T_e = 2R_e/c + T$. Obviously the two interpretations are equivalent.

Once the ranges of interest have been established, the number of samples of the analog waveform must be specified. The sampling theorem states that a continuous waveform must be sampled at a rate greater than twice its highest frequency in order to preserve all the information in the signal. Assume that the sampling frequency f_s satisfies this theorem. Then taking samples of the analog signal every $T_s = 1/f_s$ will result in a sequence containing the same information. Converting to a notation which is better suited to sequences: let $\vec{r}_{if}(n)$ be the n'th sample of the analog signal $r_{if}(t)$. Then for the N samples beginning with n equal to zero

$$
\begin{aligned}
\vec{r}_{if}(n) &= r_{if}(T_i + nT_s) \\
&= \cos[2\pi(f_i f(T_i + nT_s) + f(T_i + nT_s - \tau) + .5a(T_i + nT_s - \tau)^2)]
\end{aligned}
\tag{1.19}
$$

where N was selected as the smallest integer such that $T_i + (N-1)T_s \geq T_e$.

Recall that the matched filter for compressing the analog signal r_{if} can be implemented as a correlation, a time reversed linear filter, or a conjugate frequency domain receiver. The linear digital filter which implements the time reversed receiver can be obtained by sampling the impulse response of the corresponding analog system. The same sampling frequency that was used for obtaining $\vec{r}_{if}$ from r_{if} must be maintained. Recall that $h(t) = u^*(-t)$, where $u(t)$ was defined in Equation 1.18. Sampling $h(t)$ every $T_s = 1/f_s$ results in a sequence similar to the one shown in Figure 1.8. Because the received signal r_{if} is taken over a range T_i, T_e which must be at least as long as the duration of the pulse T, the number of samples of h will be less than or equal to the number of samples of r_{if}. Therefore use the

Impulse Response $h(t) = u^*(t)$

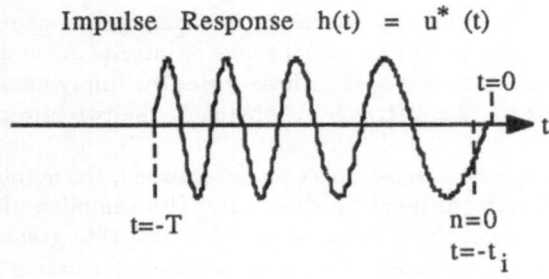

Sampled Impulse Response $h(n)$

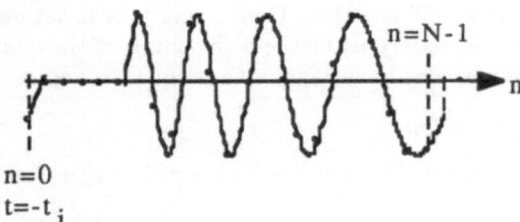

Figure 1.8: Sampled matched filter impulse response.

longer sequence length, which will always be the N for the received signal, as the length for all the digital sequences to be processed. This adjustment which makes all sequences equal length contributes several subtleties to the following steps.

We now have a sequence representing the received data and a sequence representing the impulse response of the time reversed matched filter receiver. As is explained in Appendix A, sampled data is treated as one period from an infinitely long periodic sequence. For the impulse response $h(t)$, the nonzero values occur for time $t \leq 0$. These values must be "wrapped around" to be within the first N samples of the sequence. That is, the sampled impulse response of length N is given by

$$\vec{h}(n) = \begin{cases} h\left(-t_i\right), & \text{if } n = 0; \\ h\left(-t_i + (n - N)T_s\right), & \text{else,} \end{cases} \tag{1.20}$$

where t_i equals T_i modulo T_s. This shift in the time axis by t_i is done to align the samples of the reference with the samples of the reflected signal as is depicted graphically in Figure 1.8. There are other techniques for designing digital filters from a continuous time representation which do not directly sample the impulse response function. One alternative for the discrete matched filter is to sample the continuous time reference function directly with no complex conjugate or time reversal and then perform these modifications in the frequency domain with a complex conjugate operation.

The digital matched filter described by the unit sample response $\vec{h}$ can be implemented as the discrete convolution of $\vec{r}_{if}$ with $\vec{h}$. If this operation is to be performed by circular convolution, then the length N needs to increased by padding zeros to $\vec{r}_{if}$ and $\vec{h}$. This can be thought of as padding the zeros to $\vec{h}$ before the periodic interpretation of the sequence is invoked. The use of zero padding to eliminate wrap around is very important in many applications where a circular convolution is used to perform a linear convolution. In a radar system performing matched filter pulse compression, circular convolution is sufficient because wrapped around data represents correlations between partial pulse returns which are considered useless. Suppose the length of the sampled reference pulse without appended zeros is given by N_{ref}. Recall that an entire pulse echo is required in the received data in order to have a valid matched filter. Once $N - N_{ref} + 1$ comparisons (correlations) have been performed the wrap around begins. With a padded sequence, zeros begin at this point. Regardless, there will not be a complete pulse echo available for compression. Therefore the padding of zeros is usually not performed and the wrapped around values are simply discarded. The reduction in signal length by avoiding zero padding improves performance by decreasing computation time. The output of the matched filter can therefore be calculated by the circular convolution of $\vec{h}$ with $\vec{r}_{if}$ which is denoted as $\vec{y} = \vec{r}_{if} \otimes \vec{h}$.

As is demonstrated in Appendix A, the multiplication of the Discrete Fourier Transforms (DFT) of two sequences is often a faster and more efficient calculation than performing the circular convolutions directly. Let $\vec{R}_{if}$, $\vec{H}$, $\vec{U}$, and $\vec{Y}$ be the DFTs of the sampled received signal, the sampled matched filter impulse response sequence, the sampled reference waveform, and the output of the circular convolution implementation of the matched filter, respectively. Then by the definition of the DFT

$$
\begin{aligned}
\vec{Y}(k) &= \vec{R}_{if}(k)\vec{H}(k) \tag{1.21}\\
&= \vec{R}_{if}(k)\left\{\sum_{n=0}^{N-1} \vec{h}(n)e^{-j2\pi nk/N}\right\}
\end{aligned}
$$

Recalling the definition of the unit sample response sequence $\vec{h}$ and the time reversed relationship $h(t) = u^*(-t)$ implies

$$
\begin{aligned}
\vec{Y}(k) &= \vec{R}_{if}(k)\left\{h(-t_i) + \sum_{n=1}^{N-1} h\left(-t_i + (n-N)T_s\right)e^{-j2\pi nk/N}\right\}\\
&= \vec{R}_{if}(k)\left\{u^*(t_i) + \sum_{n=1}^{N-1} u^*\left(t_i + (N-n)T_s\right)e^{-j2\pi nk/N}\right\} \tag{1.22}
\end{aligned}
$$

Letting m equal $N - n$, taking the complex conjugate operation outside of the sum, and noting that $e^{j2\pi k}$ is one for any integer k results in

$$
\begin{aligned}
\vec{Y}(k) &= \vec{R}_{if}(k)\left\{u(t_i) + \sum_{m=1}^{N-1} u(t_i + mT_s)e^{+j2\pi(N-m)k/N}\right\}^*\\
&= \vec{R}_{if}(k)\left\{\sum_{m=0}^{N-1} u(t_i + mT_s)e^{-j2\pi mk/N}\right\}^* \tag{1.23}
\end{aligned}
$$

If the reference sequence $\vec{u}$ is defined as $\vec{u}(n) = u(t_i + nT_s)$, for $0 \le n \le N-1$, and its corresponding DFT is denoted by $\vec{U}$, then the expression for $\vec{Y}$ can be rewritten as

$$
\vec{Y}(k) = \vec{R}_{if}(k)\left\{\vec{U}(k)\right\}^* = \vec{R}_{if}(k)\vec{U}^*(k). \tag{1.24}
$$

This is, of course, the discrete version of the conjugate receiver where the final sequence $\vec{y}$ is obtained by taking the inverse DFT of $\vec{Y}$. The continuous time reference signal is sampled directly (no time-reversal or complex conjugate) with a time shift of t_i to place the reference correctly between samples.

Because the DFT's can be computed using Fast Fourier Transform (FFT) algorithms, this conjugate receiver implementation of the matched filter is typically faster and more efficient in memory usage than the corresponding circular convolution time reversed discrete filter or the circular correlation receiver. When the pulse and data sequences are of substantially different lengths, however, direct implementation of the discrete convolution may be more efficient. This issue will not be pursued in these notes although it is often an important aspect of radar implementation.

1.4 Seasat Radar Processing

The primary mission of the Seasat satellite when it was put into orbit by JPL/NASA in June 1978 was to map large regions of the earth's surface, especially the oceans. Because the Seasat experiment is used in the next chapter as an example of a synthetic aperture radar (SAR) imaging system, it is also presented here to demonstrate one dimensional radar processing techniques. The Shuttle Imaging Radar (SIR) projects which are also SAR imaging systems carried on the space shuttle in 1981 (SIR-A) and 1984 (SIR-B) use very similar processing.

The Seasat radar system operated in the microwave region of the spectrum transmitting a linear fm chirp waveform. The reflected signal is received in one channel where it is beaten down (mixed) to an intermediate frequency and then sampled. The pulse and sampling parameters are presented in Table 1.1. Recall that the functional form of a linear fm transmitted pulse is given by

$$s(t) = \begin{cases} \cos[2\pi(f_c t + .5at^2)], & 0 \leq T; \\ 0, & \text{else}, \end{cases} \qquad (1.25)$$

where T is the pulse duration of 33.8 μsec which begins at time $t = 0$. It is also possible to center the pulse about the time origin. Inserting the appropriate parameters for the Seasat experiment results in

$$s(t) = \cos[2\pi(1275.t + .5(19./33.8)t^2)]\Re_T(t), \qquad (1.26)$$

Note that the units for the time variable t are μsec and the chirp rate a is given by the bandwidth of the pulse divided by the pulse duration or 19./33.8 MHz per μsec. The rectangle function $\Re_T(t)$ is zero everywhere except when t is in the interval $[0, T]$ where it is unity.

The echo, or reflected signal, can be represented as $r(t)$ which equals $\sigma s(t - \tau)$ for targets at time delay τ. Mixing the received signal down so

Table 1.1: Radar parameters for the Seasat experiment.

Nominal Altitude .... 794 km
Transmitter (Carrier) Frequency (f_c) .... 1275 MHz
Pulse Chirp Rate (a) .... .5621 MHz/μsec
Pulse Duration (T) .... 33.8 μsec
Pulse Bandwidth (aT) .... 19. MHz
Intermediate Center Frequency $(f_{if} - .5aT)$ ..... 11.38 MHz
Pulse Repetition Rate .... 1645 Hz
Sampling Rate (A/D) .... 45.03 MHz
Sampling Duration $(T_e - T_i)$ .... 288 μsec
Antenna Dimensions .... 2m by 10.5m

that the pulse is centered about the intermediate frequency results in

$$r_{if}(t) = \sigma \cos[2\pi(f_c\tau + f_{if}t + .5a(t-\tau)^2)]\Re_T(t-\tau) \qquad (1.27)$$
$$r_{if}(t) = \sigma \cos[2\pi(-1275.\tau + (11.38 - 19./2.)t + .5(19./33.8)(t-\tau)^2)].$$

Note that the role of the intermediate frequency is to center the pulse about the frequency 11.38 MHz. This is why 11.38 MHz is called the intermediate center frequency and the IF used in mixing is 11.38-19/2 MHz. An equivalent approach would be to use an IF of 11.38 MHz and transmit a pulse centered about the time origin. This would have created a 19 MHz pulse centered about the carrier frequency f_c. The signal r_{if} in Equation 1.27 is the response, after mixing, of one target and is therefore zero outside of the width of the pulse. The total response from M different ranges before mixing would be $\sum_{m=1}^{M} \sigma_m s(t - \tau_m)$. Because the processing is done using linear operations, the effect of many targets is simply the sum of the returns for each processed individually. This is a consequence of the superposition property of linear operations and allows the analysis to consider only the simpler single target case.

Knowing the form of the intermediate signal which has been received and sampled allows the design of a matched filter for compressing the pulse. For this signal r_{if} the continuous time reference waveform is

$$u(t) = \cos[2\pi(f_{if}t + .5at^2)]\Re_T(t).$$
$$= \cos[2\pi((11.38 - 19./2.)t + .5(19./33.8)t^2)]\Re_T(t). \quad (1.28)$$

This waveform must be sampled before it can be used to process $\vec{r}_{if}$. The Fortran pseudo-code given in Figure 1.9 could be used to generate the complex array necessary to use this reference waveform for the digital conjugate

```
c          Matched Filter Reference Generation
c          time units of μsec and frequency units of MHz
c          reference() = complex( real_part, imaginary_part )
c
           time_initial = 0.0
           pulse_duration = 33.8
           pulse_bandwidth = 19.0
           chirp_rate = pulse_bandwidth/pulse_duration
           sampling_frequency = 45.03
           sampling_interval = 1.0/sampling_frequency
           f_if = (sampling_frequency/4) - (pulse_bandwidth/2)
           do 100 i=1,length
               time = time_initial + (i-1)*sampling_interval
               if( time ≤ pulse_duration ) then
                   phase = 2π * (f_if*time + 0.5*chirp_rate*time²)
                   reference(i)=complex( cosine(phase), 0.0 )
               else
                   reference(i)=complex( 0.0, 0.0 )
               endif
100        continue
           call FFT(reference, length)
           call CONJUGATE(reference, length)
           call WINDOW(reference, length)
```

Figure 1.9: Fortran pseudo-code for generating matched filter reference.

receiver implementation of the matched filter. The array **reference** in this routine corresponds to the sampled reference pulse $\vec{u}$. For convenience the time shift t_i of the origin (time_initial) is set to zero. Because the FFT efficiently computes the DFT in place, **reference** becomes $\vec{U}$ after the subroutine call. The standard Fortran convention of starting arrays at position one is maintained in this pseudo-code. This generates a one position shift in the indices. That is, $\vec{U}(0)$ is stored in **reference(1)** and in general, $\vec{U}(k)$ is **reference(k+1)**. This confusing shift is common in signal processing code and should be kept in mind. The CONJUGATE routine multiplies the imaginary term of the array by minus one to perform the complex conjugate operation. This creates the matched filter coefficients for the frequency domain implementation.

The matched filter coefficients returned by routine CONJUGATE are

```
c        Matched Filter Implementation
c        Perform complex multiply between reference and data
c
         call FFT(data, length)
         do 200 i=1,length
             data(i) = data(i) * reference(i)
200      continue
         call IFFT(data, length)
```

Figure 1.10: Fortran pseudo-code for implementing matched filter.

sufficient for performing the pulse compression operation, however, a call
to subroutine WINDOW was also included for generality. The WINDOW
routine multiplies the values of the complex array **reference** by some pre-
determined weighting. Inverted cosine and other smooth bell shapes are
often used for the purpose of reducing the side lobes of the response. Side
lobes are the spikes which occur next to the center lobe of the matched
filter output (recall the autocorrelation function). Trimming the height of
the side lobes makes the central spikes easier to find by reducing bleeding
of information from one target response onto another. This improvement,
however, is accompanied by a spreading of the autocorrelation's main lobe.
There is no exact science to selecting the appropriate window for a given
data set. Seasat data is often windowed, but it is not a necessary step in
the processing. In addition, windowing, like techniques for increased speed,
should probably only be attempted after a working system exists. This
allows reasonable comparisons to be made without creating code that is
difficult to debug.

The reference array generated in this routine can be used to process
the sampled received data $\vec{r}_{if}$. If $\vec{r}_{if}$ is stored in the complex array called
data, then a simple code for performing the matched filter operation and
conversion back to the time domain is given in Figure 1.10. Note that as
long as the IFFT routine accepts inputs in the same order as the FFT
produces output, the circular convolution operation is performed. This
allows the elimination of the bit reversal operation.

The number of nonzero points in the time sequence $\vec{u}$ generated by sampling the reference $u(t)$ is approximately the pulse duration multiplied by
the sampling frequency. The **reference** array for the Seasat pulse compression therefore has 33.8 μsec times 45.03 MHz or about 1530 nonzero
points. If the length used for the FFT's is 4096, then only the first 2567

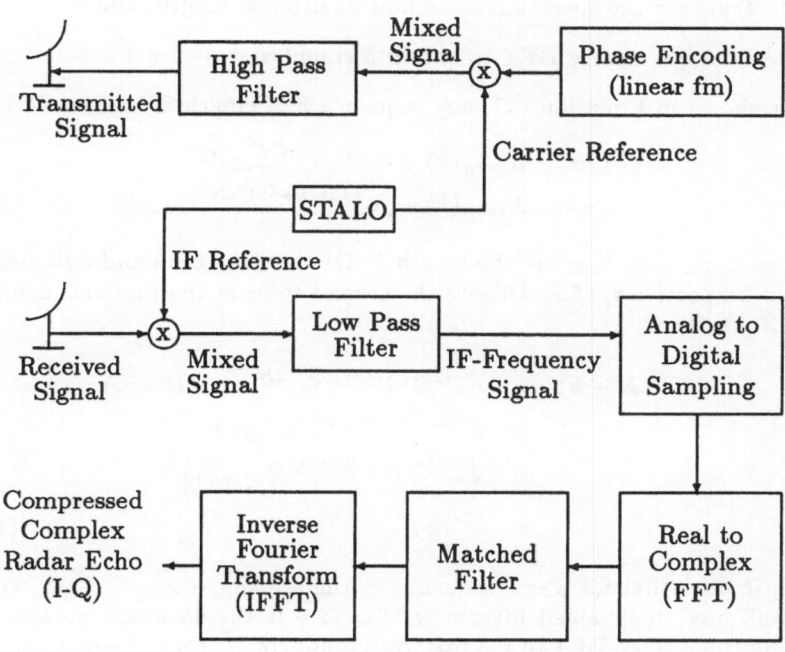

Figure 1.11: System for processing Seasat data.

points of the output **array data** are valid. The other 1529 points represent the wrapped around portion of the circular convolution and are not used in the next step of the radar image processing.

For reasons which will be explained in the following chapter, Seasat data is not processed as a real signal but as a set of complex numbers. Because the raw data was sampled in one channel the conversion from real to complex representation will be performed digitally. This digital conversion from real to complex numbers corresponds to the analog in-phase and quadrature (I-Q) systems described earlier. The complete Seasat radar system including the analog transmitter and receiver and the digital signal processing scheme is presented in Figure 1.11.

The digital real to complex conversion is usually accomplished with

FFTs. This relatively efficient procedure is

1. Take the DFT of the real sequence $\vec{x}$ to obtain $\vec{X}$;

2. Truncate the spectrum $\vec{X}$ to half its original length; and

3. Take the inverse DFT to obtain a complex sequence $\vec{Y}$.

As is shown in Equation C.1, any sequence $\vec{x}$ of length $2N$ has a DFT

$$
\begin{aligned}
\vec{X}(k) &= \vec{X}_{even}(k) + e^{-j2\pi k/2N}\vec{X}_{odd}(k) \\
&= \vec{X}_{even}(k) + je^{-j2\pi(k+\frac{N}{2})/2N}\vec{X}_{odd}(k) \quad\quad (1.29)
\end{aligned}
$$

where $\vec{X}_{even}$ and $\vec{X}_{odd}$ are the length N DFTs of the even and odd indexed parts, respectively, of $\vec{x}$. Denote the second term in the previous equation as $j\vec{Z}(k)$, then

$$
\begin{aligned}
\vec{Z}(N-k) &= e^{-j2\pi((N-k)+\frac{N}{2})/2N}\vec{X}_{odd}(N-k) \\
&= e^{+j2\pi(k+\frac{N}{2})/2N}\vec{X}_{odd}^{*}(k) \\
&= \left\{ e^{-j2\pi(k+\frac{N}{2})/2N}\vec{X}_{odd}(k) \right\}^{*} \\
&= \vec{Z}^{*}(k) \quad\quad (1.30)
\end{aligned}
$$

This implies that for a real sequence $\vec{x}$, the sequences $\vec{X}_{even}$, $\vec{X}_{odd}$, and $\vec{Z}$ will all have real-valued inverse DFTs. If y is the sequence obtained by taking the inverse DFT of the first $N/2$ points of $\vec{X}$, then the real part of $\vec{y}$ will equal the even indexed elements of $\vec{x}$ and the imaginary part of $\vec{y}$ will equal $\vec{z}$ which is a function of the odd indexed points of $\vec{x}$ as defined above. An intuitive interpretation of the imaginary terms of $\vec{y}$ can be obtained by analyzing the coefficient of $\vec{X}_{odd}$ in the definition of $\vec{Z}$. This coefficient can be separated as

$$
e^{-j2\pi(k+\frac{N}{2})/2N} = e^{-j2\pi k/2N}e^{-j\pi/2}. \quad\quad (1.31)
$$

Note that multiplication of a frequency sequence of length N by $e^{-j2\pi mk/N}$ induces a circular shift in the time series by m positions. That is, $\vec{x}_{odd}(n-m)$ has a Fourier transform equal to $e^{-j2\pi km/N}\vec{X}_{odd}(k)$. For the case at hand, m equals $1/2$ (a noninteger), which can be interpreted as a one half sample shift (interpolation) of the sequence $\vec{x}_{odd}$ to the time positions midway between samples. These new time positions correspond to the sample positions for $\vec{x}_{even}$. The effect of the second term in the coefficient is a constant one quadrant phase shift. This $\pi/2$ shift converts a cosine to a sine and completes the representation of the in-phase and quadrature signal. Note that the real and imaginary parts of the final complex sequence came from independent samples of the original sequence $\vec{x}$ and the sampling period is now doubled. In short, the information represented by two single-channel real-valued samples is equivalent to one two-channel complex-valued sample sampled at intervals twice the real-valued sampling period. There are other interpretations for the I-Q conversion; Problem 12 describes the I-Q conversion using Hilbert transforms and analytic signals.

There is one detail that has been overlooked in this discussion; the number of frequency terms required to completely specify the DFT for a real sequence of length $2N$ is $N + 1$ not N as assumed. The errant term is $\vec{X}(N)$ which for a real sampling frequency of 45.03 MHz corresponds to the 22.5 MHz frequency term of the signal. For Seasat data, no information from the received signal is contained in this specific frequency and it is therefore zeroed as noise allowing length N transforms to be used.

The results of this real to complex conversion are intuitively satisfying. The complex sequence has real part equal to samples of the real signal taken at half the original rate and a spectrum equal to the first half of the original real sequence spectrum. The processing can be performed using the same reference generation as before. The real to complex conversion of the reference is accomplished by using the first half of the spectral coefficients of the matched filter. Because the first and last step of the conversion is an FFT, the transition is incorporated as part of the matched filter processing. Figure 1.12 presents Fortran pseudo-code for implementing this complex (I-Q) radar system without increased computation for the real to complex conversions. The use of complex data in radar imaging systems makes this an important approach. Note, however, that the truncation of the spectral data for conversion to the complex representation is easily performed with the data in normal order. Elimination of the bit reversing stage of the FFTs would require additional control code.

1.5 Summary of Radar Processing

The effective duration and energy of the transmitted pulse determine the resolution and maximum range, respectively, of a radar system. Shorter duration pulses allow closely spaced targets to be discriminated, while high energy pulses provide measurable reflections from targets at large ranges. In order to avoid the difficult and expensive development of hardware to generate short duration pulses with large energy characteristics, increased duration pulses are coded for transmission and then compressed at echo reception. A linear fm (chirp) format is often used for pulse coding and a correlation (matched filter) receiver is used for compression. With pulse

```
c          Complex Matched Filter Implementation
c          Perform complex multiply between reference and data
c
           call FFT(data, length)
           do 300 i=1,(length/2)
               data(i) = data(i) * reference(i)
300        continue
           call IFFT(data, (length/2))
```

Figure 1.12: Fortran pseudo-code for implementing complex matched filter.

coding, the maximum range resolution is determined by the bandwidth of the transmitted pulse.

The phase information contained in a radar echo can be recorded if the signal is coherently received. This usually requires a stable local oscillator (STALO) which produces a reference signal to mix (multiply) with the echo. A complex number representation known as in-phase and quadrature (I-Q) simplifies the analysis and is physically represented by a receiver with two coherent channels whose references are separated by one quadrant (90 degrees) in phase. The analog processing for coherent radar reception and pulse compression can also be implemented on a computer using digital signal processing techniques.

1.6   References

M. I. Skolnik, *Intro. to Radar Systems,* McGraw-Hill Book Company, New York, 1980.

M. I. Skolnik, ed., *Radar Handbook,* McGraw-Hill Book Company, New York, 1970.

R. E. Ziemer and W. H. Tranter, *Principles of Communications,* Houghton Mifflin Company, Boston, 1976.

1.7   Problems

P1.1 Assume that rocks and sound travel with constant (but different) velocities v_{rock} and v_{sound}, respectively. What is the depth of a well having a measured round trip echo delay of time T?

P1.2 Assume that a rock falls with a constant acceleration due to gravity of g and that sound propagates with a constant velocity of v_{sound}. What is the depth of a well with a measured round trip echo delay of time T?

P1.3 Range ambiguities in radar systems and data collisions in a packet communication (computer) network represent similar problems. Given a data packet of time duration T, a propagation speed of s, and N transceiver nodes, what is the minimum time delay between transmissions to insure that no collisions from any other node will occur for all the receivers? Assume the receivers are evenly spaced in

a) a ring configuration and

b) a straight line geometry.

What effect would coding the pulse for error correction have on this system?

P1.4 For the autocorrelation $A(\tau)$ of an arbitrary complex time signal $u(t)$ prove

a) $A(\tau) = A^*(-\tau)$;

b) the Fourier transform of $A(\tau)$ is real and nonnegative; and

c) $A(0) \geq A(\tau)$, for all $\tau \neq 0$.

P1.5 Verify Equation 1.7.

P1.6 What is the autocorrelation function for a linear fm pulse with rate a, initial frequency f, and of duration $2T$ which begins at time $-T/2$ and ends at $T/2$?

P1.7 Derive the analytic forms for the Fourier transform and the autocorrelation of the following functions:

a) Rectangle of duration T and

b) One-sided exponential with decay rate a.

P1.8 Show that for complex-valued signals $u(t)$ and $v(t + \tau)$, maximizing the integral squared difference is equivalent to minimizing the real part of the complex correlation function $C(\tau) = \int u^*(t)\, v(t + \tau)\, dt$.

P1.9 Given a reference function $u(t)$, show directly that the correlation receiver $\int u^*(s)\, r(t + s)\, ds$ is linear and shift-invariant in the input signal $r(t)$.

P1.10 If the return signal $r(t)$ for a radar system is given as the superposition of target reflections $\sigma(t)$ weighting the transmitted waveform $u(t)$, or

$$r(t) = \int \sigma(\tau)\, u(t - \tau)\, d\tau,$$

and if $R(f)$, $\Sigma(f)$, and $U(f)$ are the Fourier transforms of $r(t)$, $\sigma(t)$, and $u(t)$, respectively,

a) What is $R(f)$ in terms of $\Sigma(f)$ and $U(f)$? Solve for $\Sigma(f)$ and call this $\Sigma_{IF}(f)$ for the inverse filter estimate.

b) What is the matched filter estimate $\Sigma_{MF}(f)$ of the reflectivity?

c) Why are Σ_{IF} and Σ_{MF} approximately equal for linear fm modulated signals?

P1.11 If a linear fm pulse has a large time-bandwidth product, show that the first zero crossings of its autocorrelation function $A(\tau)$ occur at τ approximately equal to $1/(aT)$. (Equation 1.8 should be useful).

P1.12 The Hilbert transform of a real-valued signal $u^r(t)$ is given by

$$u^h(t) = \frac{1}{\pi} \int_{-\infty}^{\infty} \frac{u^r(\tau)}{\tau - t}\, d\tau$$

where the integral is the Cauchy principal value (accounts for the singularity at $\tau = t$). An analytic signal $u(t)$ is defined in terms of its real-valued representation and its Hilbert transform as

$$u(t) = u^r(t) + j\, u^h(t)$$

If $U(f)$, $U^r(f)$, and $U^h(f)$ are the Fourier transforms of $u(t)$, $u^r(t)$, and $u^h(t)$, respectively, and given the identity

$$U^h(f) = -j\,\mathrm{sgn}(f)\,U^r(f) = \begin{cases} -j\,U^r(f), & \text{if } f > 0; \\ 0, & \text{if } f = 0; \\ j\,U^r(f), & \text{if } f < 0. \end{cases}$$

explain the use of FFTs for real-valued to I-Q conversion in the coding example of Figure 1.12 in terms of the Hilbert transform.

Chapter 2

Radar Imaging

An image or picture can be thought of as the values of a two dimensional function. The natural representation of an image by digital computer is as a matrix or a two dimensional array. Each element of the matrix represents the value of the image at some position specified by the row and column indices. When a matrix represents an image or picture, the elements of the matrix are usually referred to as pixels—which is a popular contraction for picture elements. The matrix (digital image) may be generated in many different ways: by sampling a function with two continuous parameters, by a functional mapping of one discrete image onto another, or by processing data into a format amenable to two dimensional display and interpretation.

A computer printer can be used to generate an image by specifying which positions on the output paper receive ink. If the size of the dots made by the printer are large, then it is difficult to generate nice looking output. The ink dots and the blanks, of course, correspond to the pixels in the image. The imaging system can be considered as a large binary matrix which determines the two dimensional distribution of the ink. A crt can be interpreted in a similar manner: the phosphor is turned on at the positions determined by the imaging system. The electron gun in the crt, however, may be able to control the intensity as well as the position of the dot. This implies that the "ink" is not binary, but rather some gray scale of values ranging from bright to dark. Understanding the process of displaying an image provides some insight into potential geometries for the collection of two dimensional data. Of obvious interest here are geometries appropriate for radar imaging systems.

There are radar systems which use focused lasers to scan target areas and record the reflectivity at incremental positions of the scan. For each position of the radar beam a single value is recorded which is representative of the reflectivity of the target at that position. These returns are stored in a matrix which can be displayed as an image on a crt with intensity proportional to the estimates of the target reflectivity. As with the printer, however, the quality of the image will be determined by the size of the pixel, or in this case, the size of the radar beam as it reflects off the target. This technique is descriptively referred to as focused spot scanning. Note that each radar range echo is condensed into a single value representing target reflectivity while the range information from the echo is ignored.

If more than one range is of interest or if the radar signal propagation can be used as a means of scanning the target area, then the range direction becomes one of the image dimensions. This implies that one continuous return or echo (possibly after pulse compression), will generate an entire column of data points. When the analog signal is appropriately sampled for digital representation the resolution in the range dimension is determined by the bandwidth (effective time duration) of the pulse. This radar format is typical of the systems described in the previous chapter. By moving the radar system and transmitting another pulse, radar reflections for a neighboring set of targets can be processed. Repeating this procedure results in a set of radar reflections which, when juxtaposed, form an image. If the transmitted radar signal is focused so that each reflection is from a narrow strip of targets, then the resolution of this system will be very good. In ultrasonic signal processing, this technique of accumulating an image by juxtaposing single echo returns is known as the B-scan (the name is a consequence of the individual echoes being referred to as A-scans). The radar discipline calls the juxtapositioning of neighboring echoes to form a composite image "strip mapping."

Both strip mapping and focused spot scanning can be used effectively in radar and other imaging systems when a sharply focused beam can be maintained. When well focused beams are not feasible, alternate means of achieving resolution are required. Synthetic Aperture Radar (SAR) is a processing technique applied after collection of the radar data which improves the resolving (focusing) power of the system. The extra information needed to reconstruct a high resolution image is obtained by illuminating the target areas many times. Recall that high resolution corresponds to small pixel sizes. Additional processing for improved resolution is typically required for spaceborne radar imaging systems where the illuminating beam diverges substantially over the propagation path but can also be used as a means of improving the resolution of lower altitude systems. The discussions which follow concentrate on the issues of spaceborne radar imaging—including the restrictions, geometries, approximations, and the resulting processing techniques which are peculiar to imaging large surface areas from high altitude sensors. As in the previous chapter, the Seasat experiment will be used as an example.

2.1 Restrictions on Antenna Size

A pulse emitted from an antenna will illuminate some target area. For radar systems which are scanning the earth's surface, the ground area one pulse illuminates is called the radar's footprint. The geometry for a typical radar and its footprint are given in Figure 2.1. The ground surface is specified by the along track (azimuth) and cross track (ground range) coordinates. Because the range (not ground range) is measured directly as the round trip propagation delay and represents the absolute distance from antenna to target, it is often a more convenient parameter than the ground range. The dimension of the footprint is a function of the antenna size, the range, and the transmitted wavelength. An antenna of dimension D at wavelength λ, and at range R, will produce a footprint of approximate width $\lambda R/D$. The footprint of the radar signal does not immediately disappear outside of this region but fades quickly—like an AM radio station out of range. The width $\rho = \lambda R/D$ specifies the size where the power of the footprint is half the maximum power. In engineering this is called the 3dB level; in optics this is called the far field or Rayleigh's resolution.

Satellites can orbit at many different altitudes but are nominally partitioned into three groups based on altitude: 0.2Mm corresponds to Low Earth Orbit (LEO) satellites, 2Mm to High Earth Orbit (HEO), and 40Mm represents Geosynchronous Earth Orbit (GEO) satellites, where one Mm is 1000km. For an orbiting microwave radar imaging system, such as Seasat, typical values of λ and R would be 23.5cm and 850km, respectively. Suppose that a square pixel $\rho = 25$m on a side will provide satisfactory resolution for this imaging system. Recall that the cross track (range) resolution is determined by the bandwidth of the transmitted pulse and does not depend on the physical antenna size. Because the image is being formed by strip mapping, the resolution requirement implies the radar footprint must be less than 25 meters wide in the along track (azimuth) direction. Using the 3dB rule with the wavelength, range, and resolution as fixed parameters, the dimension of the shortest antenna satisfying the requirements is $\lambda R/\rho$ which for this example corresponds to (23.5cm)(850km)/(25m) or about 8km. Obviously this is a prohibitively large antenna. Unfortunately, any imaging system with similar parameters and resolution requirements will also require an extremely large antenna—unless some technique for using less focused beams can be devised. In short, the goal is to synthesize an image with the resolution of a focused large aperture antenna system using the data returned from a physically small antenna mounted on a

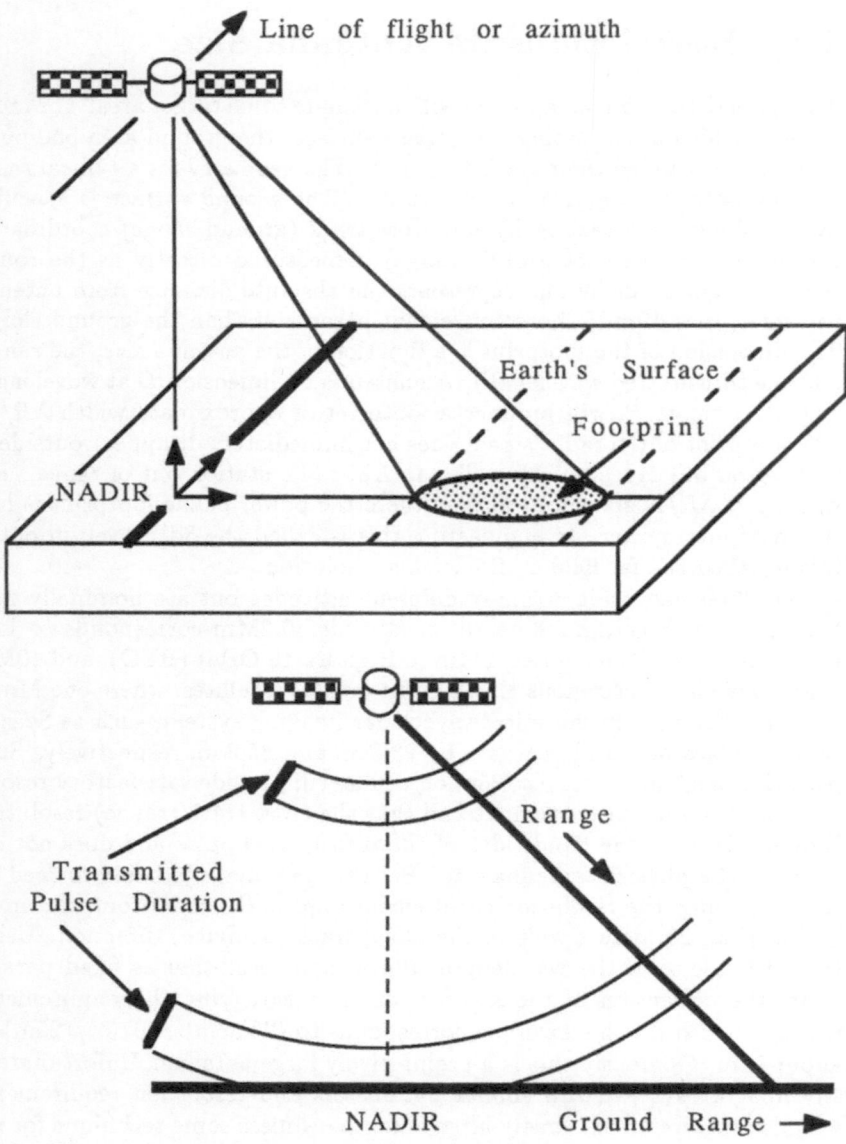

Figure 2.1: Radar geometry with transmitter and footprint.

spacecraft. Because a large antenna aperture never physically exists, the solution is known as Synthetic Aperture Radar (SAR). In the sections which follow, synthetic aperture concepts are introduced through a discussion of antenna arrays.

2.2 Antenna Arrays

Consider a simple rectangular antenna of length L and width W. An improvement in resolution (increase in focusing power for reception or decrease in beam width for transmission) can be realized by increasing the surface area or aperture of the antenna—recall the resolution is given by $\rho = \lambda R/D$. An intuitive interpretation of the relationship between antenna size and resolution will be useful in our discussion of antenna arrays. For simplicity, assume that the antenna width W is fixed by design or manufacturing requirements. Increasing the antenna dimension L results in improved focus of one dimension of the propagating wavefront. The larger antenna allows more electromagnetic radiation to be collected when receiving or emitted under transmission operation. The collected energy from every area of the antenna is channelled to a sensor for summation which results in the one dimensional time signal representing the output of the antenna. As illustrated in Figure 2.2, the incremental increase in antenna aperture could have been realized with a separate antenna performing the collection and a wire transporting the information to the central sensor for summation. Repeated application of this interpretation for a desired antenna length L results in a network of physical antennas connected by delay lines (wires) to a central summer. This type of network is called an antenna array. In fact many of the space antennas which appear to be a single aperture are actually a panel of small sensors with metallic connectors to a central processor—*i.e.*, an antenna array.

The simplest array is a series of identical antennas distributed evenly along a line with equal length wires transporting the received signals to a central circuit for summation. Note that a normally incident plane wave is uniformly distributed across all the physical antenna apertures and, because the delay induced by each wire is equal, the summation operation results in the coherent average of the incident wavefront. Because the signal resulting from a very distant point source is approximately a plane wave, a linear array with equal delays is said to be unfocused or focused to infinity.

The ability to focus an antenna array to a specific range can be accomplished through very simple modifications to the unfocused implementation. Recall that focusing in a real antenna (single physical aperture) is performed by designing an appropriately shaped antenna—a parabolic con-

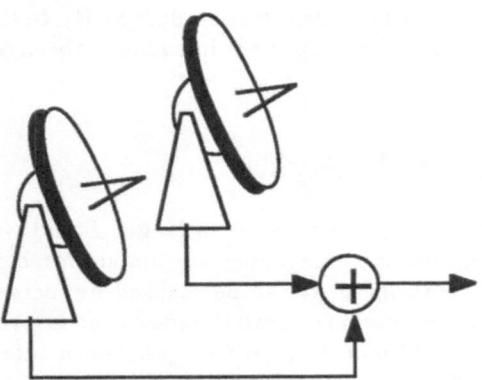

Figure 2.2: Increase of the antenna aperture.

figuration, for example. With an array, the position of individual antenna elements can be designed to mimic patterns such as parabolas. An alternative means of achieving the same effect is to incorporate delays in the wires which connect the collectors to the summers. Based on its position in the linear array, the line to each antenna is designed to incorporate a delay which produces focused reception of the wavefront. This time delay is equivalent to the propagation delay which would have been introduced by changing the physical positions of the antenna elements of the array. Figure 2.3 illustrates this concept for generating a focused array.

The problem with the focused arrays described above is that the delays incorporated in each line are functions of the focal range. That is, for fixed delay lines the array is focused for only one range. Time varying delays would allow the network to scan the focus through all ranges of interest. This can be realized with special hardware or by storing the signal provided by each delay line and then processing this set of signals into a single response. While the first interpretation presents the possibility of real time processing, the second batch type processing is more representative of the synthetic aperture radar problem. Applications like strip map imaging use the data set to generate antenna arrays focused at many different ranges. In fact, the raw data from one delay line is required by several antenna network processors in order to generate high resolution echoes for adjacent strips in the composite strip map image.

The intuitive interpretation developed to describe the reception aspects of creating a large aperture antenna from arrays of physically smaller devices provides an analogous description for transmission effects. The prob-

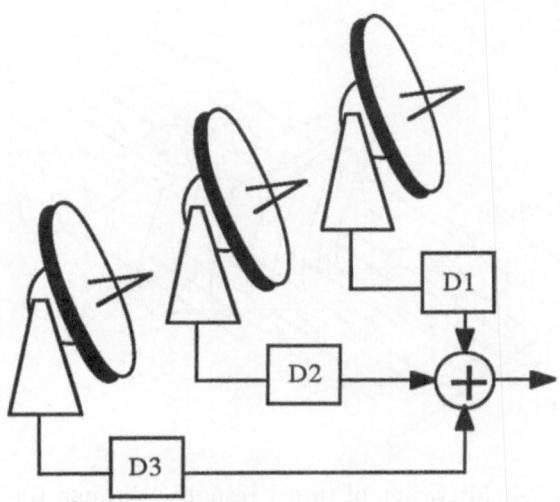

Figure 2.3: Focused antenna array.

lem of collecting a field is replaced by the problem of generating a field using an appropriate current distribution on each physical antenna of the network. In summary, high resolution strip map imaging requires well focused beams, focused beams require large antennas, and large antennas can be designed as networks of smaller antennas.

2.3   Synthetic Antenna Arrays

In airborne applications, an antenna aperture is limited by the weight and size restrictions of the aircraft. To successfully improve the performance of airborne systems, data collected by a single, physically small antenna must be processed as if it were from one element of a large antenna array. This larger aperture network does not physically exist but is synthesized by sequentially gathering data (using the small antenna) at different positions which collectively define the antenna array. This differs from storing data from each element of a real array because the transmitted radar pulse is generated as well as received by the single small antenna.

The simplest procedure for generating a sampled array requires the antenna to be sequentially positioned along a straight line for each transmit, receive, and store operation. The number of positions required to sample

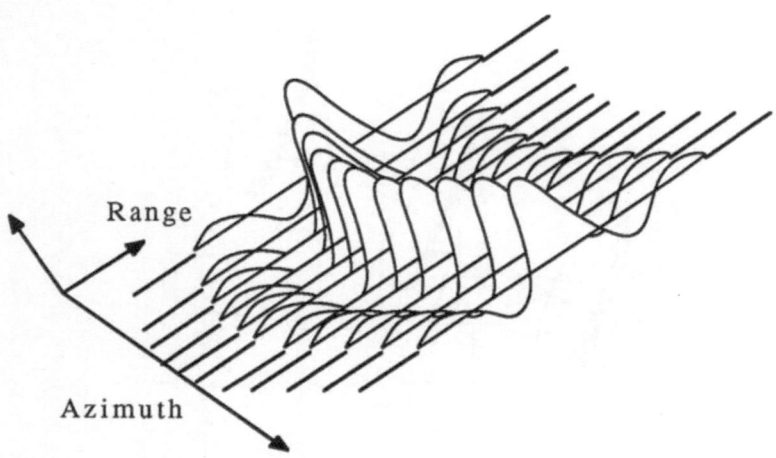

Figure 2.4: Migration of target response through the data.

the array for an arbitrary input are specified by Nyquist's theorem applied to the spatial bandwidth of the collected signal in the along track direction. An accurate time origin for the radar echo signal as well as the relative position of the antenna in the synthetic array are required for processing. Suppose that the synthetic aperture array is to be generated by averaging the stored echo signals—this corresponds to the unfocused real array previously described. Because the wavefront from the physical antenna propagates as a spherical shell, and not as a plane, there is variation in range sensed across a plane which is parallel to the array. The migration of a target response through several range positions in the time delay echo is illustrated in Figure 2.4.

The variation in range introduces relative time delay shifts among the echoes from the individual antenna positions. Averaging the echoes together at one fixed time delay will blur the signal over the region of range variation. The maximum allowed variation in range is equivalent to both the maximum allowed phase error across the plane and the maximum allowed synthetic aperture length. The relationships among these three parameters can be derived geometrically from Figure 2.5. If L is the proposed width of the synthetic aperture (which must be less than the width of the real antenna's footprint), and R is a range of interest, then the variation in

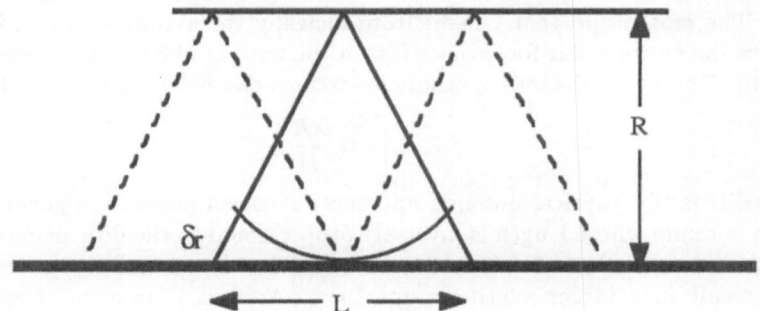

Figure 2.5: Sampled array geometry.

range δr across the aperture is given by

$$\left(\frac{L}{2}\right)^2 + R^2 = (R + \delta r)^2. \tag{2.1}$$

When the range R is much greater than δr, the expression reduces to

$$L_{unfocused} = \sqrt{8R \cdot \delta r}, \tag{2.2}$$

where the subscript has been added because this synthetic array processing scheme does not incorporate focusing.

The phase variation is related to the range variation through the wavelength λ of the radar carrier. A phase delay of ϕ degrees represents a range delay of

$$2 \cdot \delta r = \frac{\phi\lambda}{360}. \tag{2.3}$$

The δr term is doubled to account for the round trip delay induced by this variation. Therefore a maximum phase variation of 90 degrees corresponds to a maximum range variation of $\lambda/8$ or a maximum synthetic array length of

$$L_{unfocused} = \sqrt{R\lambda}. \tag{2.4}$$

Note that the maximum allowed synthetic array length will, for this type of processing, always depend on the range of interest and the broadcast wavelength.

As with real arrays, synthetic arrays benefit from focusing operations. Recall that focusing corresponds to the introduction of range dependent delays in the data set before summation. The range variation across the footprint resulting from the change in relative target distance as well as imprecise positioning of the antennas, can be accounted for during processing. The most important benefit from focusing the synthetic array, is this allows the entire radar footprint width to be used as the synthetic aperture length. Therefore, focused synthetic apertures can have a maximum length

$$L_{focused} = \frac{\lambda R}{D}, \tag{2.5}$$

where D is the physical antenna aperture. Focused processing generates a large antenna whose length is inversely proportional to the dimension of the physical antenna used to sample the array. That is, a smaller real antenna will result in a larger synthetic aperture. At first it may be surprising that a smaller physical antenna is required to increase the resolution of the synthetic array. However, a decrease in antenna size will increase the footprint width and, consequently, increase the maximum allowed length for the synthesized array.

Use of a single antenna to sample the array in both transmit and receive modes effectively doubles the phase sensitivity of the synthetic array as compared to a real array of the same length. Therefore, the resolution of the synthetic array network is

$$\rho_{sar} = \frac{\lambda R}{2L}. \tag{2.6}$$

As shown in Figure 2.6, the resolutions of a physical antenna (real array), an unfocused synthetic array with a 90 degree maximum phase variation, and a focused synthetic array are given by

$$\begin{aligned} \rho_{real} &= \frac{\lambda R}{D}, \\ \rho_{unfocused} &= \frac{\lambda R}{2L_{unfocused}} = \frac{\sqrt{\lambda R}}{2}, \\ \rho_{focused} &= \frac{\lambda R}{2L_{focused}} = \frac{D}{2}. \end{aligned} \tag{2.7}$$

The resolution of the focused system operating at maximum synthetic array length is independent of range and improves as the physical antenna dimension D is decreased!

2.4 Airborne Synthetic Arrays

In order to demonstrate and analyze the synthetic aperture principal for airborne applications, consider the azimuth-range (x, R) coordinate system

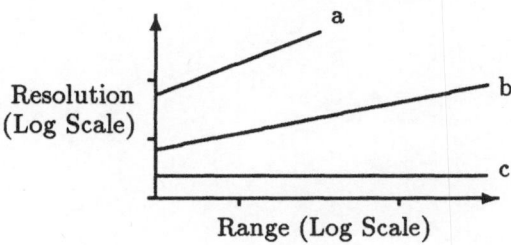

Figure 2.6: Resolution comparison for a) a physical antenna aperture, b) an unfocused synthetic aperture, and c) a focused synthetic aperture.

and point reflecting targets defined by Figure 2.7. Whenever the target at $(0, R_0)$ is within the radar footprint $(-L/2 \leq x \leq L/2)$, the sensed range as a function of the azimuthal position x of the antenna is given by the hyperbolic expression

$$R(x, R_0) = \sqrt{R_0^2 + x^2}.$$

Note that the sensed range $R(\cdot)$ equals the range coordinate R_0 only when the target is broadside—*i.e.*, when x is zero. An arbitrary target located at coordinates (x_i, R_i) contributes to the range echo through

$$R(x - x_i, R_i) = \sqrt{R_i^2 + (x - x_i)^2}, \quad \text{for} -L/2 \leq x - x_i \leq L/2.$$

This expression for the measured range $R(\cdot)$ in terms of the azimuth-range position (x_i, R_i) of a point reflector is derived using simple geometric relations which can be verified using the coordinate system defined in the Figure 2.7.

The range echo at each antenna position is usually recorded as a function of the time delay but can be converted to a range measure using the constant propagation speed c of the radar wavefront. The time delay t equals twice the range to the target divided by the speed of propagation or $2R/c$, where the factor of two accounts for round trip propagation. For targets at a known location (x_i, R_i), the relationship can be expressed in terms of the antenna position x, or

$$t(x - x_i, R_i) = \frac{2R(x - x_i, R_i)}{c} = \frac{2}{c}\sqrt{R_i^2 + (x - x_i)^2} \qquad (2.8)$$

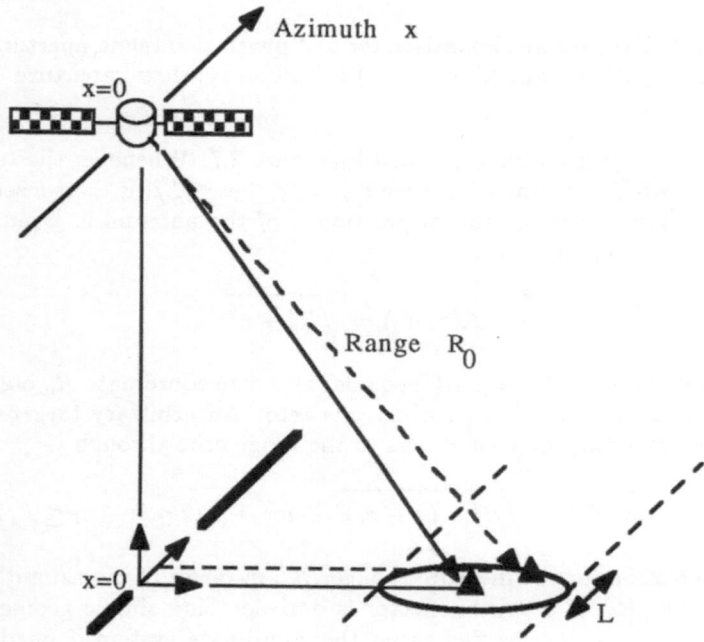

Figure 2.7: Airborne coordinate system for synthetic arrays.

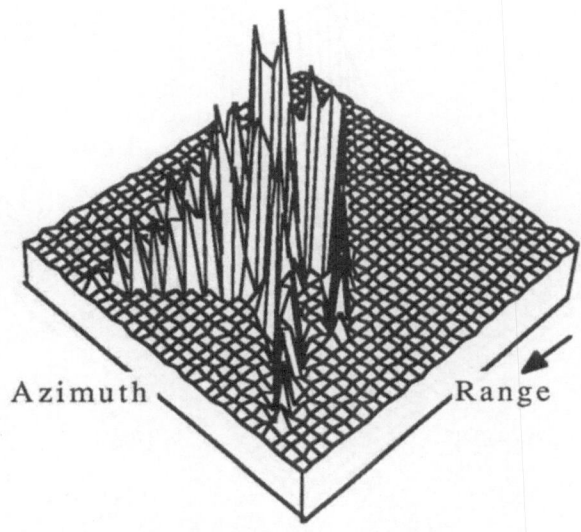

Figure 2.8: Response of point targets.

The data collected from three point targets (one pair at equal range R_0 and one pair at equal azimuth), is presented in Figure 2.8. Note that the antenna pattern weights the echoes from the point sources based on their angle of reception.

An unfocused array can be synthesized by averaging echo signals at some fixed time delay (range) bin over a window in the azimuth direction. The image resulting from processing the data from the three point reflectors is shown in Figure 2.9. The resolution is degraded by the range migration of the target echo as the summation extends along the azimuth direction away from broadside to the targets. Target information is now contained in range bins which are at a different time delay than than the normal (broadside) target range. Procedures for correcting range migration effects are known as focusing techniques and are analogous to the variable range delay lines in real antenna arrays.

In order to focus at range R_0 and along track position $x = 0$, the data is shifted (as shown in Figure 2.10) to align the target echo for summation. Computationally this corresponds to accumulating the contributions from a postulated target at (x, R_0). Note that the energy in the data due to the target at a similar range but a different azimuthal position has been distributed over several range bins. Its contribution to the (x, R_0) estimate

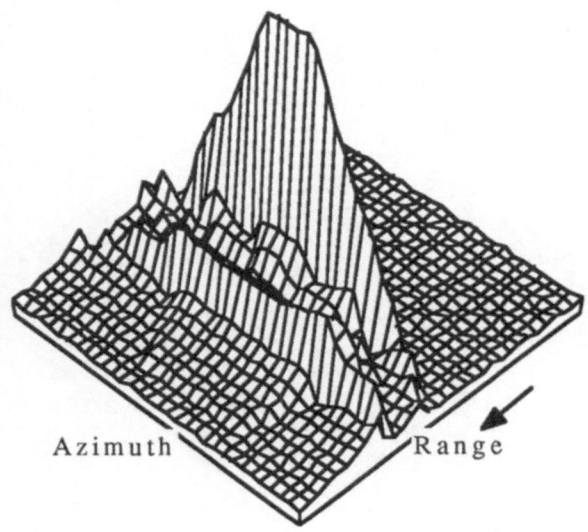

Figure 2.9: Example of unfocused synthetic array operation.

will be small after the coherent average processing introduces destructive interference from other targets. In order to generate an image focused at one range, the synthetic aperture is scanned in the azimuth direction by repeating the accumulation process with the time delays centered about each azimuth output position. This procedure produces the strip map image in Figure 2.11 which is focused at range R_0.

Although an entire two-dimensional (azimuth and range) data set results from focusing to one range R_0, there is only one range bin of interest—namely, the bin corresponding to range R_0. Simultaneous focus at all ranges represented in the time delay data set can be achieved by incorporating range dependent delays in the synthesis of the array. For each azimuth output position x, the data is incrementally focused for targets at each range. Each output range requires a particular shift in the surrounding data to create the focusing effect. Because of the hyperbolic dependance of the range migration on broadside range and azimuth position, the shifts required are different for each range. The results of using a range dependent delay for each echo contributing to the synthetic array are shown in Figure 2.12. Note that all three point targets are resolved now that the focusing is performed at all ranges and azimuth positions.

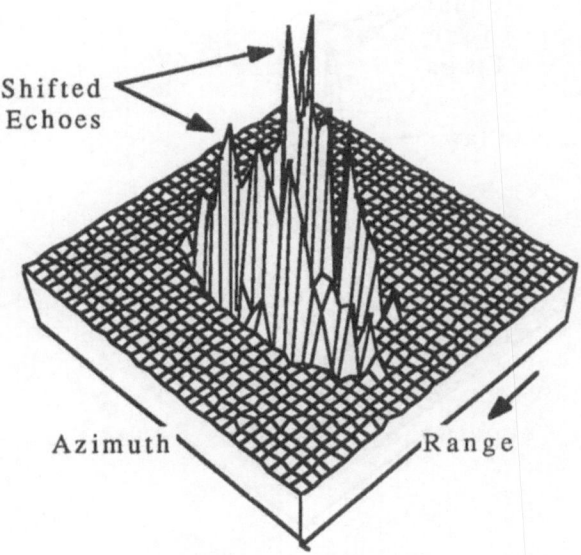

Figure 2.10: Range migration correction for target at $(0, R_0)$.

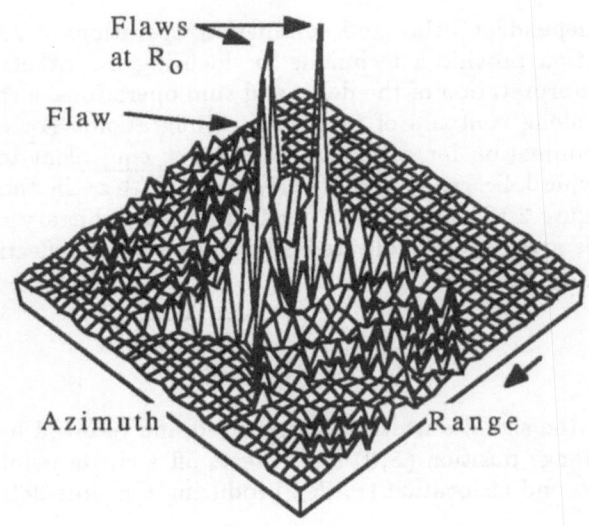

Figure 2.11: Focus at range R_0 for all azimuth positions.

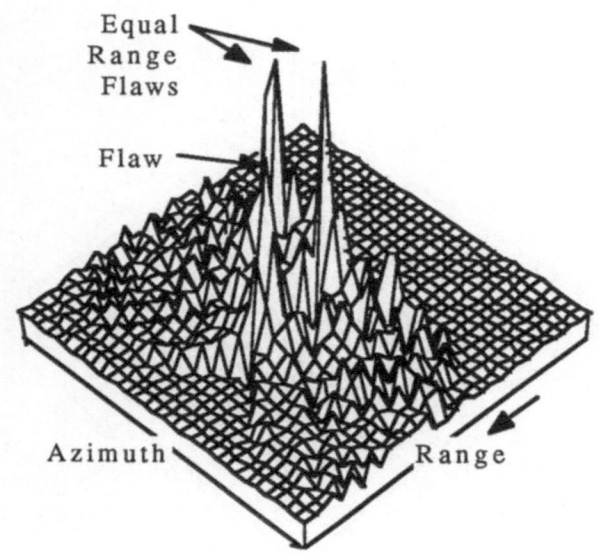

Figure 2.12: Example of focused synthetic array operation.

2.5   Matched Filter Interpretation

The range dependent delay and summation operations described in the previous section provide a technique for focusing a synthetic array. An equivalent interpretation of the delay and sum operations is as a collection of integrals along contours of target range migration. For example, the delay and summation for a target at $(0, R_0)$ is equivalent to integrating along the hyperbolic range arc $R(x, R_0) = \sqrt{R_0^2 + x^2}$ in the data set as shown in Figure 2.13. The contour for focusing an arbitrary range R_i and azimuth x_i is given by the collected echo from a point reflecting target at position (x_i, R_i) or

$$R(x - x_i, R_i) = \sqrt{R_i^2 + (x - x_i)^2}.$$

Consider the signal s which is transmitted and received by an antenna at azimuth-range position $(x, 0)$ and reflects off a single point target with reflectivity σ_i and at location (x_i, R_i) producing the time delay echo

$$r(x, t) = \sigma_i \, s[t - 2R(x - x_i, R_i)/c]. \tag{2.9}$$

The delay-sum focusing operation can be written as the integral over the

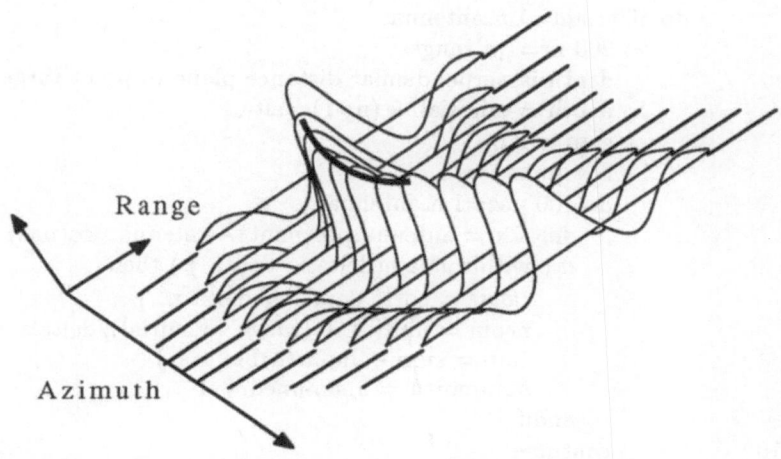

Range

Azimuth

Figure 2.13: Contour of integration for delay-sum operation.

portion of the echo data which would be produced by a point reflector at (x_i, R_i). A focused reconstruction of targets at (x_i, R_i) is given by

$$r_{focused}(x_i, R_i; t) = \int_{-L/2}^{L/2} r[x - x_i, t]\, dx \qquad (2.10)$$

$$= \int_{-L/2}^{L/2} \sigma_i\, s[t - 2R(x - x_i, R_i)/c]\, dx$$

The length L of the synthetic aperture corresponds to the width of the beam illuminating the target area (footprint). This equation can also be converted to azimuth-range (x, ρ) coordinates

$$r_{focused}(x_i, R_i; \rho) = \int_{-L/2}^{L/2} \sigma_i\, s[2(\rho - R(x - x_i, R_i))/c]\, dx \qquad (2.11)$$

In either case, these integral equations mathematically represent the linear transformation being performed by the delay and sum operation for focus at range R_i. Simultaneous focus for targets with azimuth position x_i and arbitrary range is accomplished with

$$r_{focused}(x_i; \rho) = \int_{-L/2}^{L/2} \sigma_i\, s[2(\rho - R(x - x_i, \rho))/c]\, dx \qquad (2.12)$$

```
c          Implement Focused Synthetic Array
           do 300 nant=1,n_antennas
              do 200 nr=1,n_ranges
c                 depth is perpendicular distance plane to point target
                  depth = r_initial + (nr-1)*delta_r
                  sum = 0.0
                  n_summed = 0
                  do 100 naz=1,n_antennas
                     delta_x = antenna_pos(nant) - antenna_pos(naz)
                     if ( within_beam(delta_x, depth ) ) then
                        range = sqrt( depth² + delta_x² )
                        nrbin = nint( 1 + (range - r_initial)/delta_r )
                        sum = sum + input(nrbin, naz)
                        n_summed = n_summed + 1
                     endif
100               continue
                  output(nr, nant) = sum / n_summed
200           continue
300        continue
```

Figure 2.14: Fortran pseudo-code for focused synthetic array.

For each ρ a different hyperbolic arc of data is collected (delayed) and integrated (summed). The computer code which generated the multiple range focused data in Figure 2.12 performs the integration along the expected response arcs (matched filter)—not an actual delay-sum operation. The pseudo-code for implementing simultaneous focus at all ranges is given in Figure 2.14 to demonstrate how Equation 2.12 might be implemented.

The dependence of $r_{focused}(x_i, \rho)$ on the transmitted pulse shape s is usually accomodated by positioning the data so that the maximum expected value of s is used in the integration. In the pseudo-code example, this positioning corresponds to shifting the calculation of the range by the appropriate delay in order to integrate at the position of the maximum value of the signal s rather than the first value for the appropriate range. In order to accurately estimate the target reflectivity σ, the output should be normalized by by the maximum value. Any modulation of the echo signal in the azimuth direction—e.g., antenna beam pattern weighting or phase modulation due to coherent mixing of the echo signals with a stable oscilator, can be used to improve the signal to noise ratio by incorporating a matched filter correlation of the anticipated azimuth modulation with the measured data along the appropriate range contours. Of course, more accurate tuning of references to the physical phenomena producing the measured echo will enhance performance. For this reason, the component of s along the azimuth direction and on the range migration contour is usually focused using matched filter techniques similar to the pulse compression procedures described in Chapter 1. A correlation operation is performed using the anticipated echo response as the reference to sharpen the focused response of the synthetic array.

If focus at $(0, R_0)$ is desired, a reference function of shape

$$f(x, R_0) = \begin{cases} s[2R(x, R_0)/c], & \text{if } -L/2 \leq x \leq L/2; \\ 0, & \text{else,} \end{cases} \qquad (2.13)$$

can be used to compress the radar echo in the azimuth direction. For a target at (x_i, R_0), the reference function is translated to a new azimuth position centered about x_i and expressed as $f(x - x_i, R_0)$. The shape of the reference does not change if the point target is translated in azimuth. However, a change in the desired focal range R_0 of the target will alter the shape of the reference—recall that the hyperbolic range contour $R(x, R_i)$ changes as a function of desired range focus R_i. Therefore the delay and add operation (matched filter) for focusing the synthetic array is a shift-varying operation which can be written as a convolution integral

$$r_{focused}(x_i; \rho) = \int_{-\infty}^{\infty} \int_{-\infty}^{\infty} r[x; 2R(x_i, \rho)/c] \, f(x_i - x, \rho) \, dx \, d\rho \qquad (2.14)$$

Note that the integral with respect to x is a linear shift-invariant operation, but the integral with respect to ρ is not. Focused SAR cannot be implemented using standard linear codes because fast two-dimensional correlations using Fourier transform techniques require a shift-invariant operation in both dimensions.

It is convenient to separate the two matched filtering operations and pulse compress the range echo before focusing the synthetic array. The range operations should not affect the azimuth processing because range compression effectively converts the shape of the received pulse to a more impulse shaped signal using matched filtering. The shift required to align the maximum signal response can also be incorporated in the range pulse compression operation. For radar signals recall (see Equation 1.16), that the coherently received echo signal after coherent mixing is of the form

$$r_{if}(t) = \cos\left[2\pi(-f_c\tau + f_{if}t + \phi(t - \tau))\right]\Re_T(t - \tau) \qquad (2.15)$$

where τ is the round trip time delay, T is the pulse duration incorporated using the rectangle function $\Re$, f_{if} is the intermediate frequency, f_c is the carrier frequency, and $\phi(t)$ is the pulse coding. For our previous examples, $\phi(t)$ was selected in the form $ft + 0.5at^2$ which is known as a linear fm or chirp signal. This $\phi()$ notation is often used because it is more general and compact than specifying an exact pulse shape. Recall that the $-2\pi f_c \tau$ phase term is recoverable because stable oscillators are used to generate the IF mixing sinusoids for coherent reception. Converting the dependence of the received signal r_{if} to be on the variable x, results in

$$
\begin{aligned}
r_{if}(x,t) &= \cos[2\pi(-f_c 2R(x-x_i,R_i)/c + f_{if}t + \phi(t - 2R(x-x_i,R_i)/c))] \\
&\quad \Re_T(t - 2R(x-x_i,R_i)/c).
\end{aligned}
\tag{2.16}
$$

This two dimensional signal has arguments t and x which represent round trip propagation time and azimuthal position, respectively. Recall that R_i is the normal (broadside) range to the point target at azimuth position x_i which is related to t and x by

$$
R_i^2 + x^2 = (ct/2)^2
\tag{2.17}
$$

The conversion of the one dimensional range echo to a two dimensional function is the mathematical equivalent of juxtaposing returns in strip map imaging.

Typically, this two dimensional signal reflection is converted to the complex (two channel in-phase and quadrature) representation with

$$
\begin{aligned}
r_{if}(x,t) &= e^{-j2\pi f_c 2R(x-x_i,R_i)/c} e^{j2\pi(f_{if}t + \phi(t - 2R(x-x_i,R_i)/c))} \\
&\quad \Re_T(t - 2R(x-x_i,R_i)/c).
\end{aligned}
\tag{2.18}
$$

The first phase term $e^{-j2\pi f_c 2R(x)/c}$ is usually called the azimuthal phase factor because it is not affected by the range pulse compression filter which is matched to the second term of the expression. Note that the azimuth phase factor does not depend on the time delay variable t which is equivalent to absolute range from the current antenna position. The output of the range pulse compression matched filter can be written

$$
\begin{aligned}
r_{pc}(x,t) &= e^{-j2\pi f_c 2R(x-x_i,R_i)/c} \, A_r(t - 2R(x-x_i,R_i)/c) \\
&\quad \Re_T(t - 2R(x-x_i,R_i)/c)
\end{aligned}
\tag{2.19}
$$

where $A_r(t)$ is the pulse compressed received waveform. The shape of A_r is the autocorrelation of the transmitted pulse. This is explicit in the correlation receiver implementation of the matched filter. Recall that, the autocorrelation for a complex linear fm chirp with phase $2\pi(ft + 0.5at^2)$ was calculated in Equation 1.8 as an example.

The data along the target echo curve (time delay associated with for $2R(x - x_i, R_i)/c$) can be interpreted as a one dimensional signal with phase $e^{-j2\pi f_c 2R(x - x_i, R_i)/c}$ and amplitude from A_r. The basis for synthetic aperture radar is to utilize the azimuthal phase term in the same manner as the phase coding $\phi(t)$. The phase ϕ was used to design a reference function which was matched to the transmitted signal so that information spread over time could be compressed to a narrow pulse. The azimuthal phase term now provides the shape for the reference function

$$u(x, \rho) = e^{-j2\pi f_c 2R(x, \rho)/c} \, \delta[\rho - R(x, \rho)]. \qquad (2.20)$$

When a correlation along the trace $R(x, \rho)$ is performed using this reference, compression in the azimuth direction is realized. This nonlinear trace is used so that the peak values of A_r are included in the matched filter processing resulting in a focused output with a good signal to noise ratio.

2.6 Model of the Antenna - Target Motion

An estimate of the antenna motion relative to the targets is necessary in order to perform the azimuthal compression of the signal. Errors in the antenna postion, variations from a straight-line path, and complications due to the curvature of the earth's surface will result in changes in the range arcs required to successfully focus the synthetic aperture. The trace of a point target through the echo data is usually referred to as the range migration of the system and was derived in earlier sections from the geometry of the system assuming that the antenna flight pattern is a line parallel to the target plane. In addition, there was no relative motion between the antenna and the targets—the antenna did not move between transmit and receive operations. For airborne and spaceborne synthetic aperture radar (SAR) systems this is not an accurate description of the physical radar system: the platform motion is used to scan the antenna along the synthetic aperture but must follow a nonlinear path for orbiting applications. A model which is more accurate than a line parallel to a plane is required in order to generate an azimuth reference for high resolution imaging.

The important measurement is the relative distance between the antenna platform and the target. For convenience, we define the center of the earth as the origin of the reference coordinate system. The three dimensional vectors which describe the position, velocity, and acceleration of the spacecraft are denoted by $\vec{R}_s(t)$, $\vec{V}_s(t)$, and $\vec{A}_s(t)$, respectively. The corresponding vectors for a target on the earth's surface are given by $\vec{R}_t(t)$, $\vec{V}_t(t)$, and $\vec{A}_t(t)$. These vectors are defined graphically in the coordinate

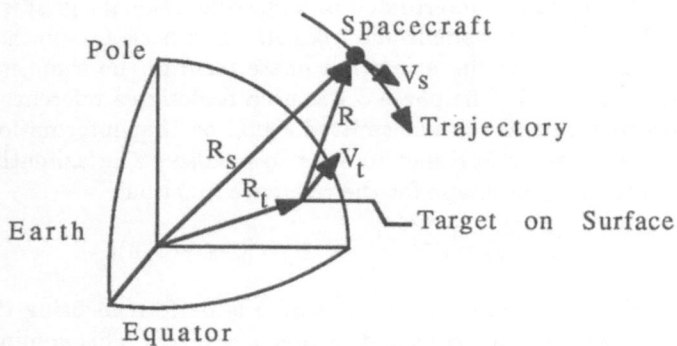

Figure 2.15: Coordinate system for the motion model.

system displayed in Figure 2.15. The important motion in the model can be described by the vectors $\vec{R}(t)$, $\vec{V}(t)$, and $\vec{A}(t)$ which are defined by

$$
\begin{aligned}
\vec{R}(t) &= \vec{R}_t(t) - \vec{R}_s(t) \\
\vec{V}(t) &= \vec{V}_t(t) - \vec{V}_s(t) \\
\vec{A}(t) &= \vec{A}_t(t) - \vec{A}_s(t).
\end{aligned}
\tag{2.21}
$$

These vectors describe the relative motion between the spacecraft and target. Note that the velocity and acceleration could also be defined in terms of the time derivative of the position vectors. When the focusing of the synthetic array (final correlation step) is implemented, an approximation to the hyperbolic range migration arcs is used because the absolute position of the spacecraft and the antenna pointing direction are not known exactly. The distances between targets and spacecraft are known in any single echo return—the change in position between echoes can only be approximated. Two different approaches for deriving the mathematical approximation used to calculate the range migration curves are presented.

For short time periods, the velocity and acceleration vectors can be considered to be constant. When the short time interval is near time t_0, the satellite to target distance can be expressed as

$$
\vec{R}(t + t_0) = \vec{R}(t_0) + \vec{V}(t_0)t + \frac{1}{2}\vec{A}(t_0)t^2.
\tag{2.22}
$$

For convenience let t_0 be zero allowing the motion to be approximated by

$$
\vec{R}(t) = \vec{R} + \vec{V}t + \frac{1}{2}\vec{A}t^2,
\tag{2.23}
$$

where the explicit time dependence of $\vec{R}$, $\vec{V}$, and $\vec{A}$ has been dropped because t is limited to a region where these vectors are constant.

The $R(x, \rho)$ variable in the azimuthal phase term equals the magnitude of $\vec{R}(t)$ for the appropriate target position vector. This magnitude will be denoted $|\vec{R}(t)|$. For small t, the magnitude squared equals

$$|\vec{R}(t)|^2 = \vec{R}(t) \cdot \vec{R}(t) \qquad (2.24)$$
$$= \vec{R} \cdot \vec{R} + 2\vec{R} \cdot \vec{V}t + (\vec{R} \cdot \vec{A} + \vec{V} \cdot \vec{V})t^2 + \vec{V} \cdot \vec{A}t^3 + \frac{1}{4}\vec{A} \cdot \vec{A}t^4,$$

where $\cdot$ denotes the inner product operation. Taking the square root of both sides results in the magnitude. The constant $|\vec{R}|$ can be factored outside of the radical, and terms associated with the third and higher powers of t can be dropped resulting in

$$|\vec{R}(t)| = |\vec{R}|\sqrt{1 + \frac{2\vec{R} \cdot \vec{V}t}{|\vec{R}|^2} + \frac{(\vec{R} \cdot \vec{A} + \vec{V} \cdot \vec{V})t^2}{|\vec{R}|^2}}. \qquad (2.25)$$

Because the the magnitude of $\vec{R}$ is typically much larger than the magnitudes of the velocity and acceleration vectors, the binomial approximation to the square root can be used to obtain

$$|\vec{R}(t)| = |\vec{R}| + \frac{\vec{R} \cdot \vec{V}t}{|\vec{R}|} + \frac{(\vec{R} \cdot \vec{A} + \vec{V} \cdot \vec{V})t^2}{2|\vec{R}|} \qquad (2.26)$$

A similar approximation to the magnitude of $\vec{R}$ can also be derived using Taylor series techniques. The required time derivatives for the second and third coefficients in the series expansion are

$$\frac{d}{dt}|\vec{R}(t)| = \frac{\vec{R}(t) \cdot \vec{V}(t)}{|\vec{R}(t)|} \qquad (2.27)$$

$$\frac{d^2}{dt^2}|\vec{R}(t)| = \frac{\vec{V}(t) \cdot \vec{V}(t)}{|\vec{R}(t)|} + \frac{\vec{R}(t) \cdot \vec{A}(t)}{|\vec{R}(t)|} - \frac{(\vec{R}(t) \cdot \vec{V}(t))^2}{(|\vec{R}(t)|)^3} \qquad (2.28)$$

Note that the velocity and acceleration vectors are defined as the first and second time derivatives, respectively, of the range vector $\vec{R}(t)$. That is,

$$\vec{V}(t) = \frac{d}{dt}\vec{R}(t) \quad \text{and} \quad \vec{A}(t) = \frac{d^2}{dt^2}\vec{R}(t) = \frac{d}{dt}\vec{V}(t). \qquad (2.29)$$

For small time variations, which we take about $t = 0$ for convenience, the Taylor series expansion is

$$|\vec{R}(t)| = |\vec{R}| + \frac{\vec{R} \cdot \vec{V}}{|\vec{R}|}t + \left\{ \frac{\vec{V} \cdot \vec{V}}{|\vec{R}|} + \frac{\vec{R} \cdot \vec{A}}{|\vec{R}|} - \frac{(\vec{R} \cdot \vec{V})^2}{|\vec{R}|^3} \right\} \frac{t^2}{2} + \cdots \qquad (2.30)$$

Because $|\vec{R}|$ is large relative to the magnitudes of the velocity and acceleration vectors, the series can be approximated by

$$|\vec{R}(t)| = |\vec{R}| + \frac{\vec{R} \cdot \vec{V}}{|\vec{R}|} t + \left(\frac{\vec{V} \cdot \vec{V}}{|\vec{R}|} + \frac{\vec{R} \cdot \vec{A}}{|\vec{R}|} \right) \frac{t^2}{2}, \qquad (2.31)$$

This is identical to the result obtained in Equation 2.26 which was derived using a second order $(\vec{R}, \vec{V}, \vec{A})$ motion model.

Recall that the azimuthal phase term depended on the value of $R(x, \rho)$, where x corresponds to the azimuthal position of the antenna platform relative to the desired focal point at depth (broadside range) ρ. Equation 2.31 implies that, for small time periods about $t = 0$, the change in range can be described by a second order polynomial. In other words, the range migration of a point target is approximated by a parabola in the echo return. By performing the second deconvolution operation for focusing the synthetic array (range compression already completed), along the appropriate parabolic curves, the maximum response for the matched filter will be obtained.

The reference function for the along track compression is given by the azimuthal phase term of Equation 2.20 which is repeated below

$$u(x, \rho) = e^{-j2\pi 2R(x,\rho)/\lambda_c} \, \delta[\rho - R(x, \rho)],$$

with $\lambda_c = c/f_c$ as the wavelength of the transmitted radar carrier. The appropriate phase ψ for the reference is therefore given by

$$\begin{aligned} \psi(t) &= 2\pi \left(2|\vec{R}(t)|/\lambda_c \right) \\ &\approx \psi(0) + 2\pi \left(f_D t + \frac{1}{2} \dot{f}_D t^2 \right) \end{aligned} \qquad (2.32)$$

where

$$\psi(0) = 4\pi|\vec{R}|, \quad f_D = \frac{2\vec{R} \cdot \vec{V}}{\lambda_c |\vec{R}|}, \quad \text{and} \quad \dot{f}_D = \frac{2(|\vec{V}| + \vec{R} \cdot \vec{A})}{\lambda_c |\vec{R}|} \qquad (2.33)$$

or

$$u(x, \rho) = e^{j\psi(0)} \, e^{j2\pi(f_D t + .5\dot{f}_D t^2)}. \qquad (2.34)$$

The f_D term is commonly called the Doppler frequency because it physically corresponds to the phase shift due to the radial velocity (the component of the spacecraft velocity in the direction of the target-antenna vector $\vec{R}$). The $\dot{f}_D$ term is usually called the Doppler frequency rate. This is because $\dot{f}_D$ is approximately the first derivative of the Doppler frequency f_D (the term weighted by $1/|\vec{R}|^3$ in Equation 2.28 would make the Doppler frequency rate exact). The relative size of $|\vec{R}|$ allows this term to be omitted without affecting the implementation. Actually this mathematical assumption or approximation does not influence the implementation of most SAR processors because the parameters f_D and $\dot{f}_D$ are estimated directly, not in terms of $\vec{R}(t)$ or its derivatives. Note that the final azimuth reference has the form of a linear fm chirp with initial frequency f_D and rate $\dot{f}_D$ which implies that the azimuth pulse compression is essentially the same operation as the range pulse compression. The range migration associated with this second order model is usually described as a combination of the linear range walk effect represented by f_D and the quadratic range curvature represented by $\dot{f}_D$.

In summary, a model of the antenna platform motion is presented and used to show that the antenna to target phase history can be approximated by a second order polynomial which allows the correlation in the along track direction to be performed in a manner analogous to the range pulse compression. Although the synthetic aperture focusing can be accomplished without relative motion between antenna and targets (knowledge of the range migration $\vec{R}$ is sufficient), the physical interpretation of the coefficients of the first and second powers of the range migration as expansion as the Doppler frequency and the Doppler frequency rate, respectively, is important in spacebourne systems where these effects dominate.

2.7   Doppler Frequency Shift

The analysis of radar signals presented in Chapter 1 centered around processing time delay information in the recorded echo. Frequency domain techniques were used for fast implementations of matched filters but not specifically for analyzing the data. If the reflected radar signals were accurately analyzed in the frequency domain, a shift which has been ignored in previous descriptions would be observed. The frequency shift is caused by the relative motion between the antenna and the reflecting target. These types of frequency shifts are known as Doppler shifts.

The Doppler shift is observed whenever a frequency generator is in motion relative to the detector. A common example of the Doppler shift is the fall in pitch of a train's whistle as it passes an observer. To better understand the source of the Doppler shift, consider the situation illustrated in Figure 2.16—a transmitter moving at velocity v_r towards a receiver. If the transmitter is sending a signal with frequency f_0, the receiver will record a signal with a slightly different frequency. For electro-magnetic waves, the speed of light can be used to convert from frequency to wavelengths. A

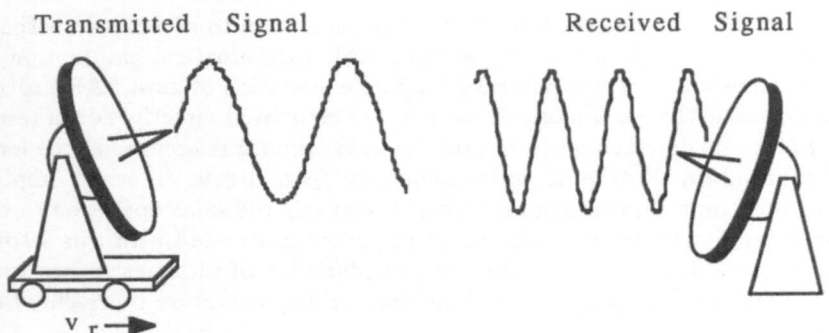

Figure 2.16: Doppler shift from relative motion between source and detector.

frequency of f_0 becomes a wavelength λ_0 equal to c/f_0. The observed shift in frequency at the receiver occurs because as a single period of the signal is broadcast, the transmitter moves a distance v_r/f_0 towards the receiver. The receiver therefore measures a distance of $\lambda_0 - v_r/f_0$ between peaks of the received signal—$i.e.$, a signal with wavelength λ equal to $\lambda_0 - v_r/f_0$. Converting this wavelength expression to frequency results in a received signal with frequency f where

$$ f = \frac{c}{\lambda} = \left(1 + \frac{v_r}{c - v_r} \right) f_0 = f_0 + f_D \qquad (2.35) $$

and f_D is defined as the Doppler frequency shift $v_r f_0/(c - v_r)$.

For radar signals the wave propagates at the speed of light c. When sound waves are used c is replaced by the speed of sound v_{sound}. Suppose that a loud noise source like an aircraft engine is generating noise at some wavelength λ_0 and is moving at velocity v_r. The Doppler phenomenon implies that a receiver on the ground would observe sound waves with wavelength $\lambda_0(1 - v_r/v_{sound})$. When the aircraft velocity v_r approaches the speed of sound, the wavelength of the sound heard by an observer tends towards zero. This stacking of large amounts of noise energy with no distance between peaks (zero wavelength) results in a burst of noise known as the sonic boom—the Doppler shift is not just some mathematical coincidence that occurs when an FFT is taken, but is founded in the basic physical laws of propagating waves.

The radar platform used in transmitting and receiving SAR signals has some velocity relative to the reflecting targets in the footprint. Because radar signals propagate at the speed of light, which is much faster than the radar platform can travel, the Doppler shifts do not create a zero wavelength signal or boom. Instead, the different radial velocities will create a continuum of Doppler shifts. The contours of targets with equal Doppler shifts (equal radial velocity), can be determined analytically. A transmitter traveling with velocity v and positioned a distance h above the NADIR point will have a velocity component in the direction of the targets in the (x, y) plane below. This radial velocity is given by

$$v_r = v \, \cos(\gamma) \tag{2.36}$$

where

$$\cos(\gamma) = \frac{x}{\sqrt{x^2 + y^2 + h^2}} \tag{2.37}$$

and γ is the angle between the line of flight and the range vector as specified in Figure 2.17.

The contours of equal Doppler shift are specified by the contours of equal radial velocity v_r. If a constant transmitter velocity v is assumed, the target positions in the (x, y) plane which imply a constant radial velocity will be specified by a constant $\cos(\gamma)$. This is an intuitively satisfying conclusion; the contours of equal radial velocity should depend only on the angle between target and transmitter relative to the direction of constant motion of the antenna. Of course this intuition is easily justified by the preceding equations.

The shape of contours of equal angle γ will specify the equal Doppler contours. This implies that the description of all points (x, y) in the plane with the same $\cos(\gamma)$ will induce identical Doppler shifts. To investigate the mathematical description of these contours, fix $\cos(\gamma)$ as a constant and rearrange Equation 2.37 to obtain

$$x^2 \cos^2 \gamma + y^2 \cos^2 \gamma + h^2 \cos^2 \gamma = x^2. \tag{2.38}$$

The trigonometric identity $\cos^2 \gamma + \sin^2 \gamma = 1$ can be used to express this as

$$\frac{x^2}{(h/\tan \gamma)^2} - \frac{y^2}{h^2} = 1. \tag{2.39}$$

This is the equation for a hyperbola with center at the origin $x = 0$, $y = 0$; distance $h/\cos \gamma$ from the center to the focus; and asymptotes of slope $\pm \tan \gamma$.

An overhead (from the transmitting antenna) view of the curves of equal Doppler shift is presented in Figure 2.18. These contours are the hyperbolas described in Equation 2.39. The smaller the angle γ, the larger the Doppler

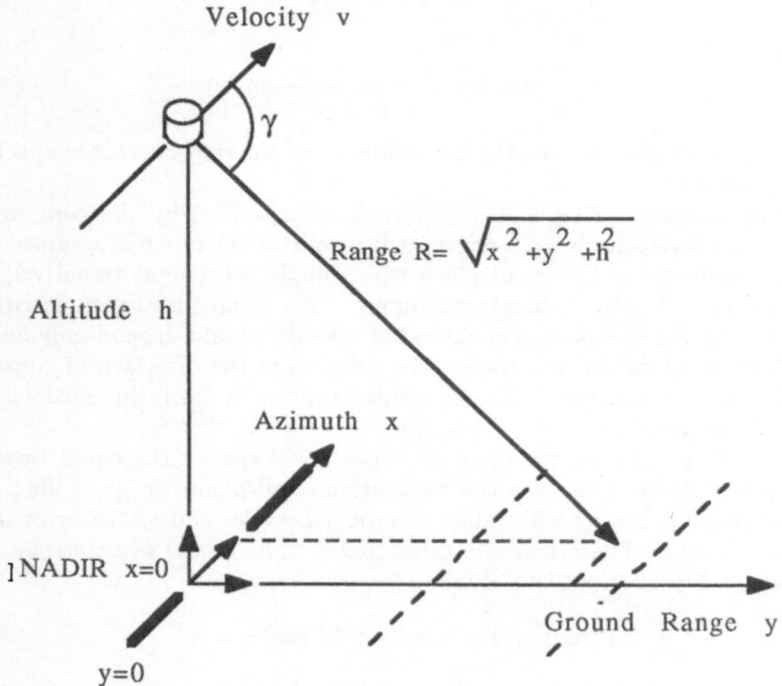

Figure 2.17: Radial velocity as a function of viewing angle.

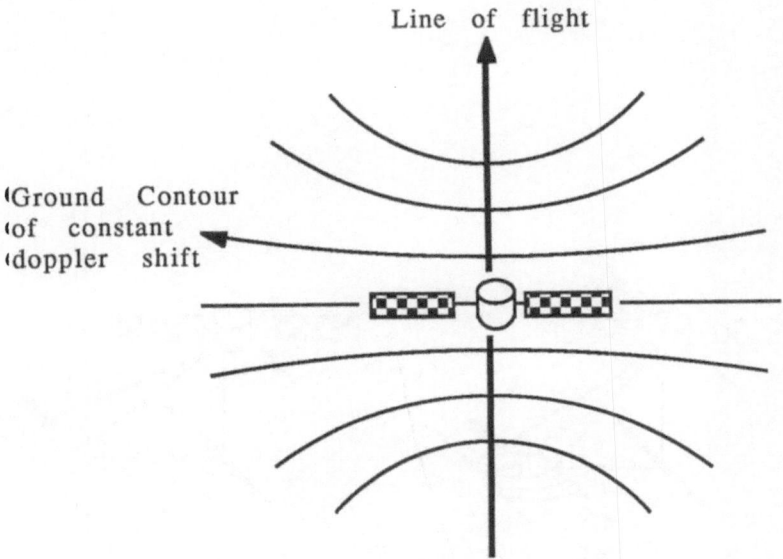

Figure 2.18: Curves of equal Doppler.

shift. Note that targets which are orthogonal to the antenna's line of flight have a zero Doppler shift. That is, targets along the y-axis have no radial velocity component. The Doppler shifts for targets in front of the satellite cause an increase in the original frequency (decrease in wavelength), targets which trail the satellite have a decreased frequency (increase in wavelength) at observation.

Although it is mathematically more convenient, it is not necessary to center the data about a Doppler shift of zero. Pointing the antenna away from broadside is called "squint mode" and is a very common technique which does not substantially change the processing from that of a normal strip map SAR. As seen in Figure 2.18, the antenna footprint must be kept to one side of the line-of-flight in order to avoid left-right ambiguities which cannot be deciphered using range-Doppler techniques. A variation of the squint mode SAR concept is spotlight mode SAR which involves steering the antenna such that the same ground patch is illuminated by every pulse in the synthetic aperture. Spotlight SAR requires different processing techniques which will be outlined in Chapter 4. A summary of the side-looking SAR geometries is given in Figure 2.19.

Values for typical Doppler shifts caused by the relative motion between

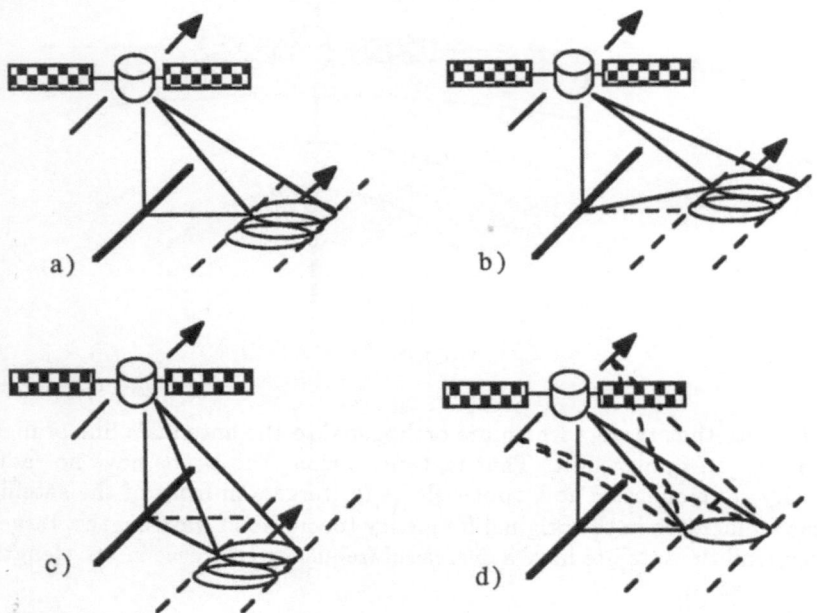

Figure 2.19: Side-looking SAR geometries: a) Normal beam; b) Forward-looking squint mode; c) Backward-looking squint mode; and d) Spotlight mode.

the satellite and targets on the earth in the Seasat experiment can be calculated. Using a nominal altitude of 800km, a view angle of twenty degrees from the antenna NADIR line to the center of the beam, and a footprint width of 15km creates Doppler frequency shifts ranging from −550 to +550 Hz. These shifts were calculated in the center of the beam. Slightly different results could be expected elsewhere in the footprint. Note that the recorded Doppler shift is twice the shift described by Equation 2.35 to account for reception by the moving antenna. This can be thought of as equal shifts at reflection and reception.

If the Doppler shift can be calculated from radar signals, then the relative velocity between target and antenna can be estimated. Police radar systems use Doppler analysis techniuqes to assess the speed of cars traveling on the highway. For spaceborne applications, the Doppler shift can be used to segment the footprint into sections of different radial velocity. Because the contours of equal Doppler lie predominantly in the cross track direction, these contours can be used to improve the along track resolution. This result can be verified mathematically by noting that the Doppler shift is a natural part of the azimuthal phase factor described in the preceding section.

Doppler analysis is an important part of most conventional radar systems where the Doppler shift provides a means of measuring the velocity of a target without destroying the pulse echo timing which provides the range or distance information. Because of its historic role in radar, many technical reports and publications describe SAR from a Doppler frequency point of view. The important contribution of Doppler phenomena to SAR is as the largest terms (Doppler and Doppler rate) in a series expansion of the range migration which defines the integration arcs (matched filter) for azimuthal focusing. The accurate estimation of the range migration arcs is the critical aspect of the azimuthal focusing—not the existence of relative motion between antenna and targets (physical Doppler shift). Focusing can be achieved even when the antenna does not move during transmit-receive operations.

2.8   Digital Implementation Considerations

Sampling requirements, matched filter implementation, and multi-look processing represent the three most common problem areas for implementing a SAR system. The sampling rates must satisfy Nyquist's Theorem in both dimensions, the matched filter reference changes with range, and the final imagery is often degraded by speckle noise or deviations in the phase information.

2.8.1   Along-Track Sampling Requirements

The pulse repetition frequency (PRF) controls the extent of the overlap of adjacent radar footprints. If the PRF is too slow, then there will not be a sufficient number of data points with information about the same target. This implies that the along track compression cannot be performed effectively and the resolution of the system remains the width of the radar footprint. This can also be interpreted as a sampling problem where the PRF specifies the sampling rate in the along track direction. If the PRF is too slow, then aliasing occurs in the along track direction. As is described in Appendix A, aliasing typically represents a non-recoverable error. To avoid aliasing, the PRF is increased above the Nyquist rate.

If the pulse repetition frequency is too high, then echoes from different pulses can overlap at the receiving antenna. This occurs when the round trip propagation time for the longest range of interest is greater than the time between transmissions plus the time delay for the shortest range. This problem is diagrammed in Figure 2.20. Let f_{prf} be the pulse repetition frequency, T be the pulse duration, c be the speed of light, and R_{far} and R_{near} be the largest and smallest ranges allowed for reflecting targets, respectively. The condition on the PRF which eliminates interfering pulses is

$$\frac{1}{f_{prf}} + \frac{2R_{near}}{c} > \frac{2R_{far}}{c} + T. \tag{2.40}$$

Rearranging terms, this inequality becomes

$$f_{prf} < \frac{c}{2(R_{far} - R_{near}) + cT}. \tag{2.41}$$

Perhaps this is better expressed as $1/f_{prf}$ which equals the time between pulse transmissions and is restricted by

$$\frac{1}{f_{prf}} > \frac{2}{c}\left(R_{far} - R_{near}\right) + T. \tag{2.42}$$

An interesting point about this condition is that only the pulse duration T and the difference $(R_{far} - R_{near})$ are needed to determine the bound on the PRF. The actual round trip time delay is not needed. This is exploited in many high altitude radar systems where several pulses are transmitted before the echo from the first pulse returns.

As an example, consider the Seasat experiment. The nominal altitude of the satellite is 840km which implies an approximate round trip time delay of 28msec. However, the pulse has a time duration of 33.8μsec and the ranges of interest have $R_{far} - R_{near}$ approximately equal to 120km. This

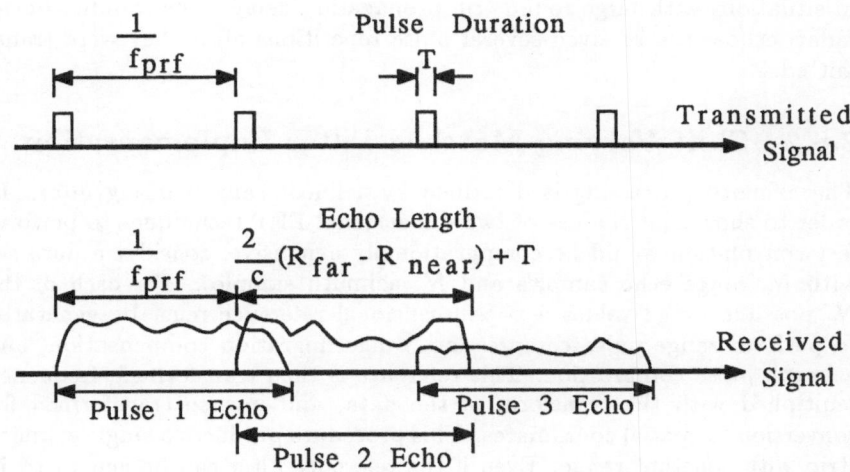

Figure 2.20: Pulse repetition frequency requirements.

implies, by Equation 2.42, that $1/f_{prf}$ should be about .8msec. Only the last .8msec of the 28msec waited for an echo will actually contain reflected radar data. An efficient means of increasing the PRF without decreasing the .8msec receiving interval is to receive echoes between transmissions. The time delay from transmission of the pulse until reception of its echo begins for the Seasat satellite is given by

$$t_1 = 9/f_{prf} + (N_t + 4)/(64 f_{prf}). \qquad (2.43)$$

The $9/f_{prf}$ term implies that at least nine pulse transmissions have occurred since the echo which is currently being received was transmitted. This term was determined by the nominal altitude of the satellite. The parameter N_t is adjusted by ground control to compensate for the altitude variations that could not be predicted in determining the first term.

The frequency content of the azimuth signal depends on the target distribution and the bandwidth of the Doppler shift. For the Seasat experiment this bandwidth is approximately 1100 Hz which requires a minimum sampling rate of 2200 Hz. The pulse repetition frequency used, however, was $f_{prf} = 1647$ Hz. This does not violate Nyquist's Theorem because the data is converted from single channel to I-Q (complex) representation which effectively doubles the sampling rate.

In summary, the PRF is selected to preserve the integrity of the along track data. The PRF must therefore be large enough to satisfy sampling criteria but not so large that echoes from different pulses can interfere. In situations with large round trip propagation delays, like satellite borne radar, echoes are received several pulse repetitions after they were transmitted.

2.8.2   Shift-Varying Matched Filter Implementation

The azimuth processing is described by a linear range-varying filter. In order to show that the use of two-dimensional FFT techniques to perform fast convolution would be computationally expensive, consider a data set with N_r range echo samples and N_a azimuth samples. For each of the N_r possible range values a two-dimensional reference must be generated to perform range pulse compression, range migration compensation, and azimuth pulse compression. This reference is then transformed, frequency multiplied with the transform of the data, and inverse transformed for conversion to spatial coordinates. This procedure produces a single azimuth strip with constant range. Even if the reference filter can be generated in the frequency domain, it is still expensive to perform N_r two-dimensional inverse FFTs on N_a by N_r images.

A more efficient approach which also uses FFTs is to sequentially perform the range pulse compression and azimuth compression operations. By shifting the range echoes and the range reference pulse, a more accurate implementation of the motion model can be performed without increased computation. In addition, the azimuth phase term is more accurately represented in the data when precision motion models are incorporated. After pulse compression and range migration compensation (sometimes called walk correction because the linear migration term is dominant at this stage), the azimuth reference is localized to a small subset of the N_r range bins. The N_r two-dimensional filter references are now of dimension N_a by N_m, where the number of migration cells N_m is smaller than N_r. The FFT implementation strategy is outlined in Figure 2.21.

A second strategy for implementing the shift varying filter utilizes one dimensional processing techniques. The range pulse compression and initial migration compensation are performed on the echoes before azimuth processing. The azimuth data is collected along the residual range migration arc for a particular range and then compressed. The azimuth matched filter is implemented as a convolution or frequency multiply. Although both filtering operations (range and azimuth) are one-dimensional, the azimuth data collection along the range migration curve requires interpolation to maintain a uniform sampling grid. The interpolation operation is computationally expensive, but necessary, because without interpolation generation of the matched azimuth filter reference is difficult and imprecise.

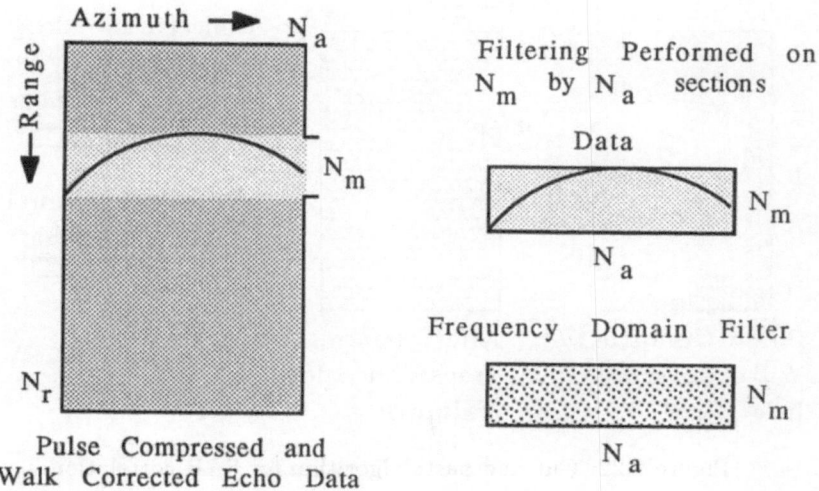

Figure 2.21: Fourier implementation of shift-varying filter.

The implementation scheme which we selected to process the data for the examples is the cut-and-paste algorithm. Range processing requires walk correction performed on the raw echo data by adjusting the starting range bin followed by range pulse compression with the appropriate matched filter. Along-track processing begins by Fourier transforming the range compressed data set in one-dimension in the azimuth direction and then collecting the appropriate signal along the residual range migration arc in frequency. Recall that the range migration is modeled as a second order process which implies that the azimuth frequency migration is also a second order (linear fm) format. A nearest neighbor interpolation can be used for collecting the data along the frequency arc for reference filter multiplication and inverse transformation to spatial coordinates—all one-dimensional operations. As shown in Figure 2.22, this algorithm cuts selected frequency regions from the azimuth data and pastes together a signal for compression.

2.8.3 Multi-Look Processing

For extremely long synthetic apertures (several thousand meters for Seasat) it is difficult to maintain the stable phase references and relative motion

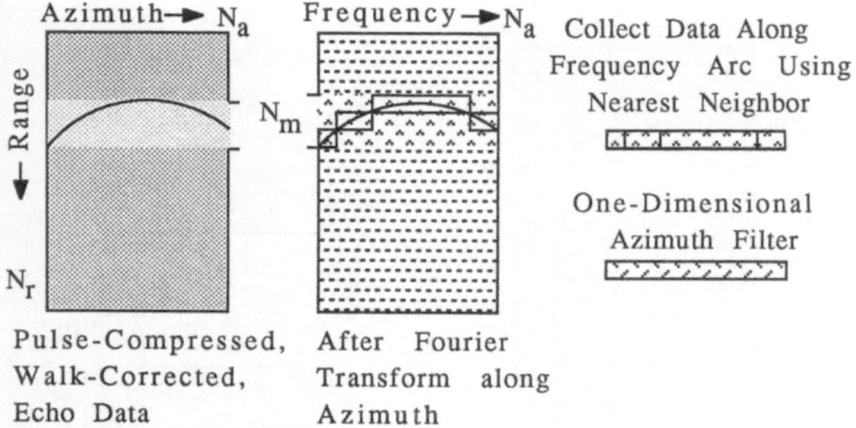

Figure 2.22: Cut and paste algorithm for SAR correlation.

measurements required for accurate azimuth pulse compression. Even if the coherence of the system can be maintained, there is an inherent noise process known as speckle which degrades image quality in any coherent system. Speckle noise is due to the variation in echo phase delay caused by targets with range variations differing by less than a wavelength. These variations create localized destructive and constructive interference which appears in the image as bright and dark speckles.

A common approach for reducing speckle is to average several independent estimates of the image. In SAR, this can be accomplished by using independent data sets to estimate the same ground patch. When only one data set is available, multiple-look filtering can be incorporated by decomposing the maximum synthetic aperture into subapertures which generate independent looks at target areas based on the angular position of the targets. For convenience assume that four looks are used: Figure 2.23 illustrates the contribution of a point target to various satellite positions and the angular positions of targets in each of the four looks. Because the satellite to target angle determines the radial velocity, the different looks can also be interpreted as different Doppler frequency bands. Note that due to the overlap required to get the same number of views of each subaperture of the footprint, twice the number of looks minus one independent reconstructions are required to complete the final multi-look image. For a four-look image, seven single-look reconstructions are required.

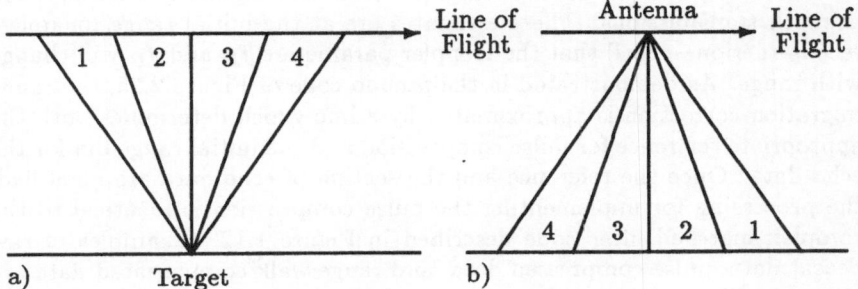

Figure 2.23: Multi-look processing: a) contribution of a point target to various satellite positions and b) angular position of targets in each of the four looks.

2.9 Seasat Image Reconstruction

A software correlator using a four-look, cut-and-paste algorithm for Seasat SAR processing was developed for use on the Mark IIA computer (a super-computer designed at Lawrence Livermore National Laboratory under the S-1 project). An outline of the processing steps is given in Figure 2.24. For each antenna (azimuth) position, a starting range sample is selected from the 13,680 four bit echo values to correct for range walk and the next 4096 values are taken as the data of interest and decoded from disk into floating point representation. The length 4096 range data is pulse compressed with a length 1529 matched filter reference; the first 2048 linearly convolved samples are converted in the frequency domain to an I-Q signal and stored in the range buffer as 1024 complex values. Four different matched filter references for the range pulse compression are used—this corresponds to using the four values of t_initial obtained when n_ref equals 1 through n_references (which is 4 in our example) in

$$t\_initial = (n\_ref - 1) * sampling\_interval / n\_references \qquad (2.44)$$

by the reference generator given in Figure 2.28. The number of consecutive range echoes which can use the same reference is given by the absolute value of

$$\frac{f_{prf} * \text{sampling\_interval}}{f_{D_0} * \lambda_c * \text{n\_references}} \qquad (2.45)$$

where $f_{D_0} * \lambda_c$ corresponds to the radial velocity v_r and v_r/f_{prf} is the estimated distance the antenna migrates from a straight-line path between azimuth transmissions. These estimates are at the initial range for image reconstruction—recall that the Doppler parameters f_D and $\dot{f}_D$ will change with range. As demonstrated in the pseudo-code of Figure 2.25, the range migration correction is approximated by a line which determines both the appropriate reference for pulse compression and the initial range bin for the echo data. Once the reference and the section of echo data are identified, the processing for implementing the pulse compression is identical to the complex matched filter code described in Figure 1.12. Examples of raw Seasat data, pulse compressed data, and range walk compensated data are shown in Figures 2.29—2.31. For our example imagery, eight 1K by 1K range blocks are processed before azimuth processing begins.

Two adjacent range blocks are used as input to the azimuth correlation processor in order to generate each of the seven single-look images required for look summation to reduce speckle. The pseudo-code in Figure 2.26 outlines the major operations necessary to perform the multi-look azimuthal SAR processing once the appropriate reference signals have been computed and the data collected along the range migration arcs. The operation corresponds to four matched filters operating on selected regions of the azimuth spectral data. In order to use this routine, the azimuth signals of two adjacent range blocks are combined to form one length 2048 azimuth signal which is converted to the frequency domain by FFT. Several of these frequency domain azimuth signals are collected (using the cut-and-paste method) into a vector for filtering by a set of four references which perform the multi-look operation. Cut points are initially described in the space domain using the range migration model and then converted to the frequency domain where the collection (paste) is performed as in Figure 2.27. The references are generated using the values for the Doppler parameters f_D and $\dot{f}_D$ associated with the current range to describe a phase history for the appropriate point focus—these operations are described by the pseudo-code in Figure 2.28. Although the filter references change continuously as a function of range (implying that the Doppler parameters should be updated with each range bin being processed), the azimuth references are kept constant over several ranges (64) in this example in order to reduce computation.

The output of the azimuth filters are four length 512 signals which, after inverse transformation to the space domain, form one azimuth line of the

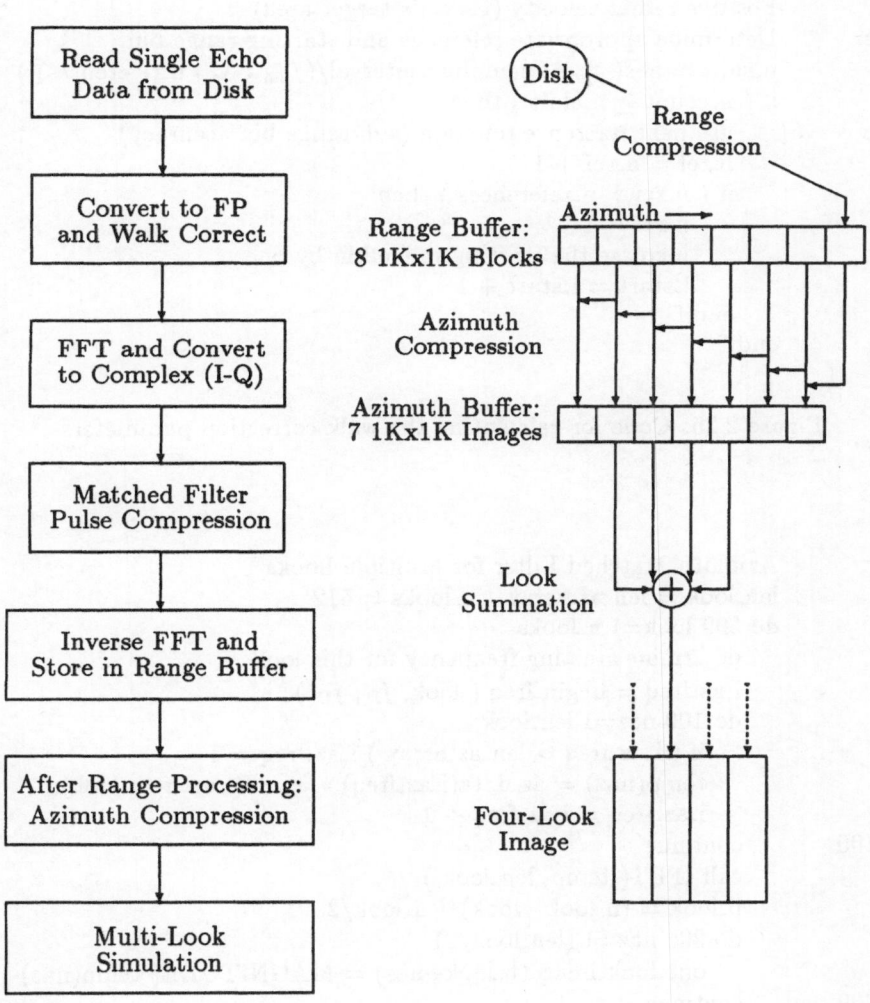

Figure 2.24: Outline of SAR processor software.

```
c       Range Walk Correction for positive f_{D_0}
c       Positive radial velocity (towards target area)
c       Determine appropriate reference and starting range bin
        n_shift = abs( f_{prf} * sampling_interval/(f_{D_0} * λ_c * n_references) )
        if ( n_count ≥ n_shift ) then
c           use next reference function (sub range bin accuracy)
            n_ref = n_ref + 1
            if ( n_ref > n_references ) then
                n_ref = 1
c               increase the starting range bin by one
                i_start = i_start + 1
            endif
        endif
```

Figure 2.25: Code for calculating the walk correction parameters.

```
c       Azimuth Matched Filter for Multiple Looks
        len_look = len_az_array / n_looks = 512
        do 300 look=1,n_looks
c           determine starting frequency for this look
            i_az_freq = begin_freq ( look, f_D, ḟ_D )
            do 100 naz=1,len_look
                if ( i_az_freq > len_az_array ) i_az_freq = 1
                temp(naz) = az_data(i_az_freq) * az_ref(i_az_freq, look)
                i_az_freq = i_az_freq + 1
100         continue
            call IFFT( temp, len_look )
            b_look = (n_look - look)*len_look/2
            do 200 naz=1,(len_look/2)
                one_look_image(b_look+naz) = MAGNITUDE( temp(naz) )
200         continue
300     continue
```

Figure 2.26: Code for multi-look azimuth pulse compression.

```
c          Collection of Data Along Frequency Domain Migration Arcs
c          First point in synthetic array always a cut
           n_cuts = 1
           t = - len_az_ref * n_looks / (2 * f_prf) = -2048 / f_prf
           f = f_D + ḟ_D t
           i_f_cut(1) = modulo( f, f_prf ) * len_az_array / f_prf
           idist(1) = -λ[(f_D - f_D₀)*t + 0.5*ḟ_D * t² ] / (2 * δr)
           do 100 naz=1,(len_az_ref*n_looks - 1)
               t = t + (1.0 / f_prf)
               i_range = -λ[(f_D - f_D₀)*t + 0.5*ḟ_D * t²] / (2 * δr)
               if ( i_range ≠ idist(n_cuts) ) then
c                  range has migrated to a different bin
                   n_cuts = n_cuts + 1
                   idist(n_cuts) = i_range
c                  convert time cut to frequency cut point index
                   f = f_D + ḟ_D t
                   i_f_cut(n_cuts) = modulo( f, f_prf ) * len_az_array / f_prf
               endif
100        continue
c          Last point in synthetic aperture always a cut
           n_cuts = n_cuts + 1
           t = len_az_ref * n_looks / 2.0
           f = f_D + ḟ_D t
           i_f_cut(n_cuts) = modulo( f, f_prf ) * len_az_array / f_prf
```

Figure 2.27: Code for determining frequency domain cut positions.

```
c          Generate Reference for Azimuth Matched Filter
           len_az_ref = n_echoes_in_range_block = 1024
           len_az_array = 2 * len_az_ref = 2048
           t = len_az_ref * n_looks / (2 *f_prf)
           do 300 look=1,n_looks
              do 100 naz=1,len_az_array
                 if ( naz ≤ len_az_ref ) then
                    phase = 2π (f_D*t + 0.5*ḟ_D * t² )
                    temp_ref(naz) = cmplx( cos(phase), sin(phase) )
                    t = t + (1.0/f_prf)
                 else
c                   zero pad for linear convolution
                    temp_ref(naz) = cmplx( 0.0, 0.0 )
                 endif
100           continue
              call FFT( temp_ref, len_az_array )
              call CONJUGATE( temp_ref, len_az_array )
              call WINDOW( temp_ref, len_az_array )
              do 200 naz=1,len_az_array
                 az_ref(naz, look) = temp_ref(naz)
200           continue
300     continue
```

Figure 2.28: Code for multi-look azimuthal reference generation.

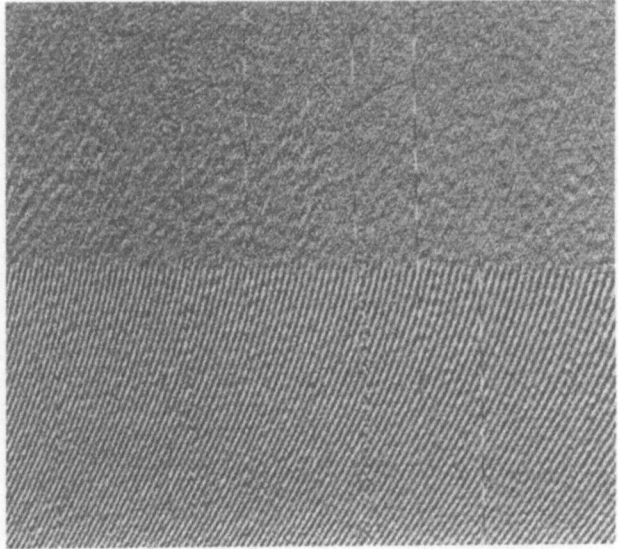

Figure 2.29: Raw data from the Seasat experiment.

four bands of a single-look image. Because the synthetic aperture has been decomposed into four angular looks, the effect of the antenna weighting pattern on the intensity of the received signal can be seen in the four bands of each single-look image (see Figures 2.32–2.38). Overlapping sections of the seven single-look images are averaged to create the final four-look image in Figure 2.39. Each quarter of the final image is the average of four single-look images of the same ground area.

The final scene is a 1024 by 1024 pixel image with each pixel representing an area on the surface of the Earth about 25m on a side. The bright star-shaped reflection in the left side of the image is a 26m antenna installed at the Goldstone Deep Space Network tracking station in southern California. The tails of the star shape, which are caused by ringing in the point response, can be reduced using windowing techniques applied in the frequency domain. The results of using a cosine-squared window are shown in Figure 2.40. The faint cross pattern of bright spots, visible to the right of the antenna, corresponds to a set of nine reflectors, ranging in size from 2 to 3m and spaced about 300m apart.

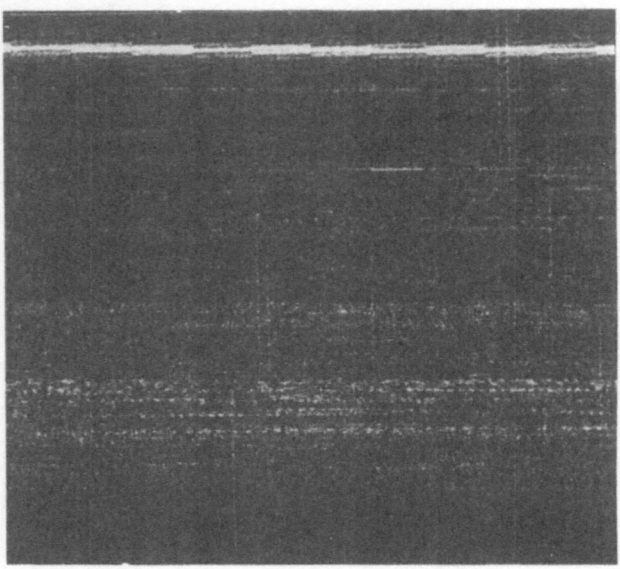

Figure 2.30: Pulse compressed data with no range walk correction.

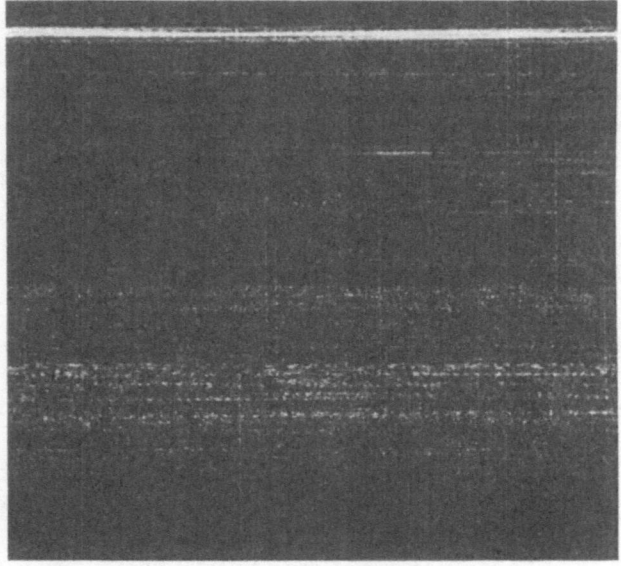

Figure 2.31: Pulse compressed data with range walk correction.

Figure 2.32: First single-look image of Goldstone area.

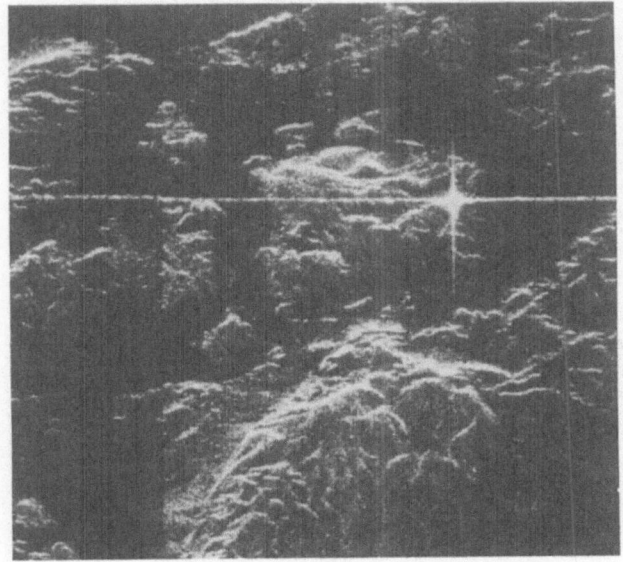

Figure 2.33: Second single-look image of Goldstone area.

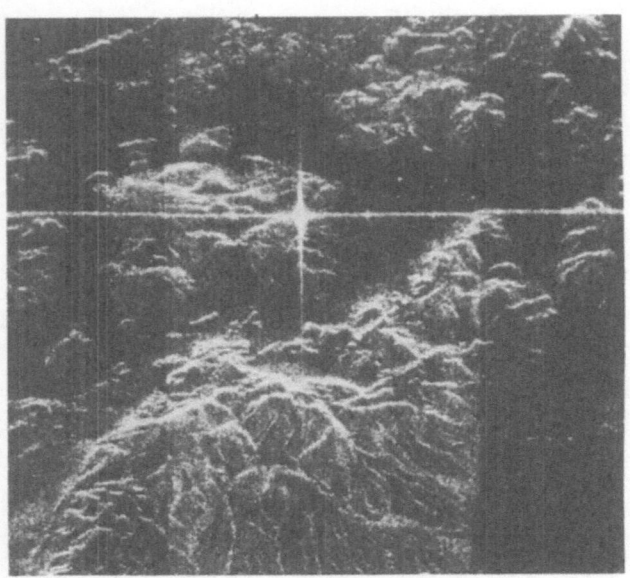

Figure 2.34: Third single-look image of Goldstone area.

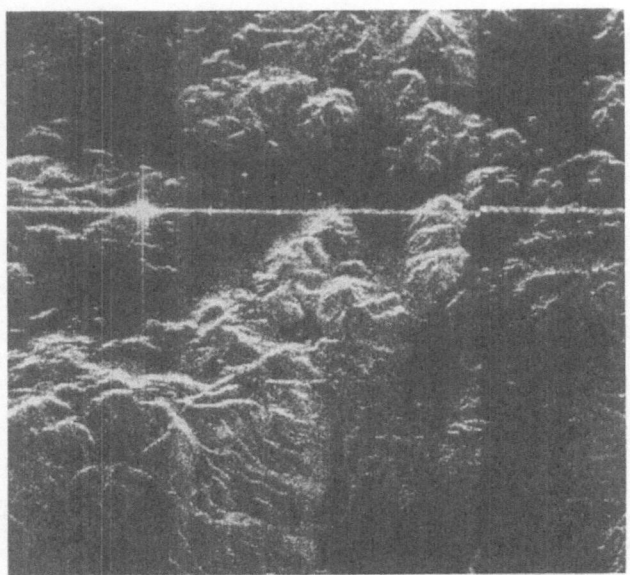

Figure 2.35: Fourth single-look image of Goldstone area.

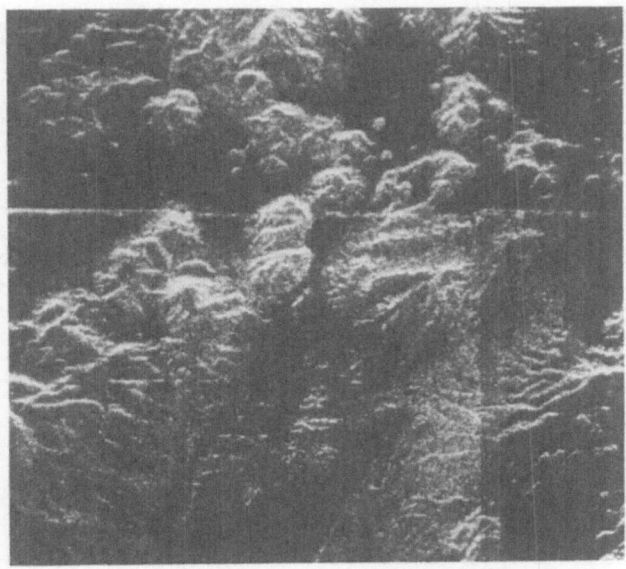

Figure 2.36: Fifth single-look image of Goldstone area.

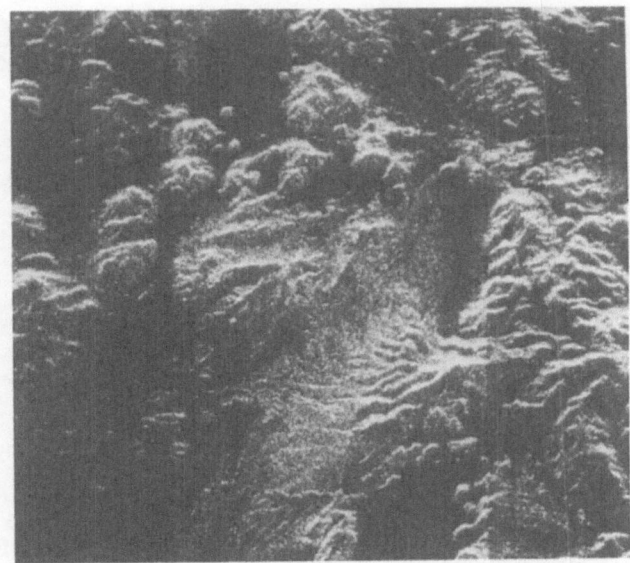

Figure 2.37: Sixth single-look image of Goldstone area.

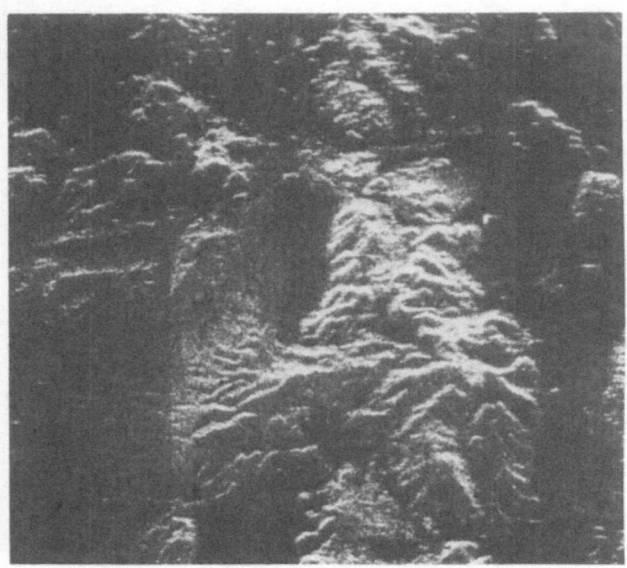

Figure 2.38: Seventh single-look image of Goldstone area.

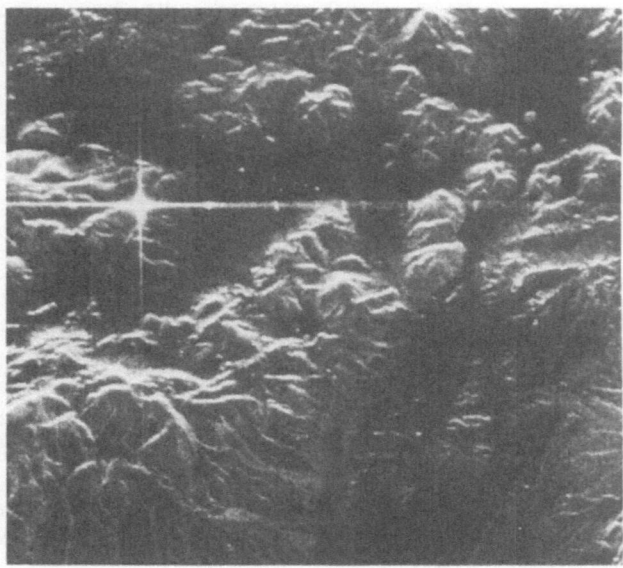

Figure 2.39: Final four-look image of Goldstone area.

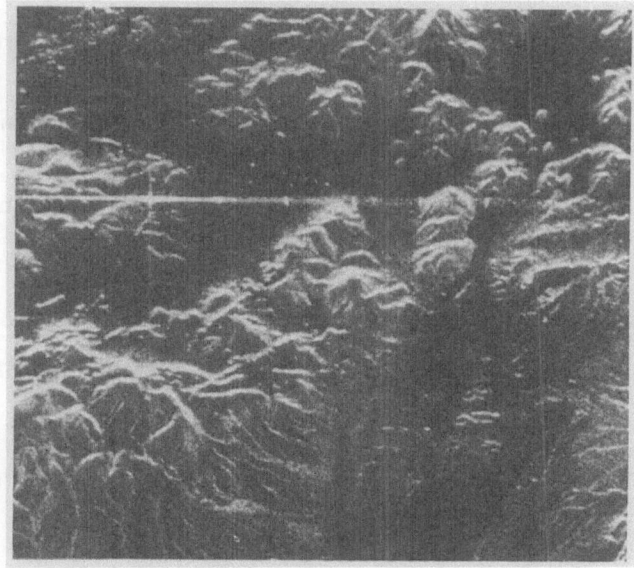

Figure 2.40: Enhancements due to windowing the Goldstone data.

2.10 Summary of Radar Imaging

Radar images can be generated using focused spots or strip maps. Synthetic arrays allow the resolution associated with a large physical antenna to be realized in systems which require the use of a small antenna. Although the processing must incorporate a shift-varying impulse response, airborne synthetic apertures can be used to generate high resolution imagery. This processing can be interpreted as either large aperture antenna synthesis or as Doppler frequency partitioning of the juxtaposed echo signals. The results are incredible: 25m resolution imagery from the Seasat satellite at a range of over 800km.

2.11 References

C. Elachi, "Radar images of the earth from space," *Scientific American*, vol. 247, no. 6, pp. 54–61, Dec. 1982.

J. P. Fitch, "Generating radar images of the earth with the S-1 computer," *Energy and Technology Review*, UCRL-52000-85-12, pp. 21–26, Dec. 1985.

R. O. Harger, *Synthetic Aperture Radar Systems, Theory and Design*, Academic Press, New York, 1970.

S. A. Hovanessian, *Intro. to Synthetic Array and Imaging Radars*, Artech House, Inc., MA, 1980.

R. L. Jordan, "The Seasat-A synthetic aperture radar system," *IEEE Journal of Oceanic Engineering*, vol. OE-5, no. 2, pp. 154–164, April 1980.

J. J. Kovaly, *Synthetic Aperture Radar*, Artech House, Inc., MA, 1976.

D. L. Mensa, *High Resolution Radar Imaging*, Artech House, Inc., MA, 1981.

C. Wu, K. Y. Liu, and M. Jin, "Modeling and correlation algorithm for spaceborne SAR signals," *IEEE Trans. Aerospace and Electronic Systems*, vol. AES-18, no. 5, pp. 563–575, Sept. 1982.

2.12   Problems

P2.1 Derive the exact expression for the length of the synthetic array L in terms of the range R and the range variation δr assuming the geometry of a straight line, averaging, sampled array (see Equation 2.1) and verify Equation 2.2.

P2.2 What is the relationship between range variation and wavelength in an unfocused sampled array if the maximum phase variation θ is given in radians?

P2.3 What is the resolution of an unfocused sampled array restricted to a maximum phase error of $\pi/16$ radians?

P2.4 Assuming that the full synthetic aperture length is used, what is the range variation in terms of wavelength λ, nominal range R, and antenna dimension D?

P2.5 What would the maximum synthetic array length for the Seasat experiment be if coherence at the STALO can be maintained over the entire footprint? What range variation δr and phase delay ϕ does this imply?

P2.6 Write the Fortran pseudo-code for calculating the walk correction parameters when f_{D_0} is negative (negative radial velocity). See Figure 2.25 for the $f_{D_0} > 0.0$ case.

P2.7 Write the Fortran pseudo-code for determining the initial azimuth look frequency for Doppler parameters f_D and $\dot{f}_D$. This corresponds to the function begin_freq in Figure 2.26.

Chapter 3

Optical Processing Of SAR Data

The concept of Synthetic Aperture Radar (SAR) is usually credited to Carl Wiley of the Goodyear Aircraft Corporation for his work in the early 1950's. Groups at both Goodyear and the University of Illinois were pursuing Doppler beam sharpening techniques around this time and an experimental verification of the concept using electronic circuitry was performed by the Illinois group in 1953. The amount of data and the type of processing required for implementing the beam sharpening taxed the capabilities of existing electronic hardware which inspired extensive research into alternate processing techniques. In 1953, a summer group at the University of Michigan (known as project Wolverine) began studying the possibility of using coherent optics for reconstruction of SAR imagery. For the next three decades, the coherent optics approach was the primary SAR processing technique. Even with todays fast digital processors, optical systems compress a large portion of the collected SAR data. In addition, photographic film provides a reasonably compact medium for storing the immense amount of raw and processed SAR data.

The following discussion begins by summarizing optical signal processing techniques and then converts digital processing functions for SAR into their optical counterparts. Because the Fourier transform is a fundamental part of both digital and optical linear systems, its properties are exploited in both domains. In order to make the discussion tractable to readers without an optics background, we will assume that the antenna position is known and that the range migration is negligible. The complications required to implement a more realistic system can be realized in optical hardware, but these details are omitted in order to emphasize the fundamental physical properties of a SAR system. After presenting a lens system which uses Fourier transforms to process SAR data, the system is modified to reduce the number of optical elements. The final system allows the imaging algorithm to be described as a wavefront reconstruction problem. This permits interpretations of SAR as holography or zone plate imaging and provides insight into the fundamental physics of the problem.

3.1 Optical Signal Processing

Perhaps the most familiar type of optical system is the imaging system. Magnifying glasses and eyeglasses are both examples of single lens imaging systems. An object a distance d_o from a lens of focal length f will be imaged a distance d_i from the lens if

$$\frac{1}{d_o} + \frac{1}{d_i} = \frac{1}{f} \tag{3.1}$$

This relationship, which is known as the imaging equation, is depicted in Figure 3.1. Using the standard optics convention, light travels from left to right in the figures. Note that many important parameters, such as the magnification $m = -d_i/d_o$, can be derived from the ray tracing techniques of geometrical optics. The magnification is negative when the image orientation is inverted relative to the object orientation. There are conventions for assigning the correct sign to the distances and the focal length which will not be discussed here. Using techniques from geometrical optics, extremely complicated lens systems can be analyzed and characterized in terms of a theoretical optical system with only a few cardinal parameters. Geometrical optics has provided a useful theory for describing imaging systems.

The typical functions of an imaging system are magnification, rotation, and translation of two-dimensional scenes. Synthetic aperture radar processing, however, requires more sophisticated operations which are often called information processing techniques. For these types of systems the important imaging relationship is replaced by the Fourier transform property of lenses. Simply stated, the distribution of light in the front focal plane of a lens is the Fourier transform of the light distribution in the back focal plane. The two focal planes are defined as planes which are orthogonal to the optical or lens axis and placed a distance f from the lens. This relationship is illustrated in Figure 3.2. If the light distribution $\psi(x,y)$ appears in the front focal plane P_f, then the light distribution $\Psi(x,y)$ appears in the back focal plane P_b, where the mathematical relationship between

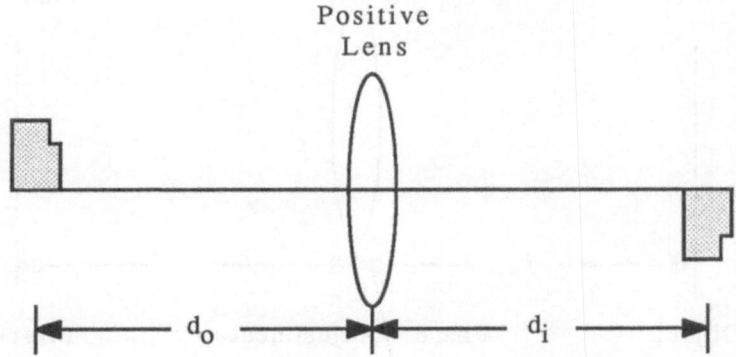

Figure 3.1: Optical imaging system.

these two distributions is the two-dimensional Fourier transform or

$$\Psi(x_b, y_b) = A \int_{-\infty}^{\infty} \int_{-\infty}^{\infty} \psi(x_f, y_f)\, e^{-j\frac{2\pi}{\lambda f}(x_f x_b + y_f y_b)}\, dx_f\, dy_f, \qquad (3.2)$$

where A is defined as the complex weighting $(j\lambda f)^{-1}$. The optical Fourier transform is an extremely parallel operation (each position in the back focal plane represents a two-dimensional transformation of the input plane) which is performed at the speed of light.

In order to use the Fourier transform property of a lens for data processing, control of the light distribution in the front focal plane is required. A specific distribution can be realized by fabricating a film transparency with a controlled transmission function $t(x, y)$. Illumination of this transparency by a spatial light distribution $\phi(x, y)$ transforms the light distribution into $\psi(x, y) = t(x, y)\phi(x, y)$. That is, the output distribution is the point-wise product of the incident light distribution with the transmission of the film. Because the transparency is not a source of energy, the transmission function's magnitude is between zero and one, *i.e*, $0 \le |t(x, y)| \le 1$. Note that because amplitude transmissions (real valued transparencies) are essentially photographic slides, they are relatively easy to fabricate. Complex (phase and amplitude) transparencies can be implemented for many special cases but are typically more difficult because the phase implementation often requires varying the thickness of the film or glass substrate which carries the amplitude information.

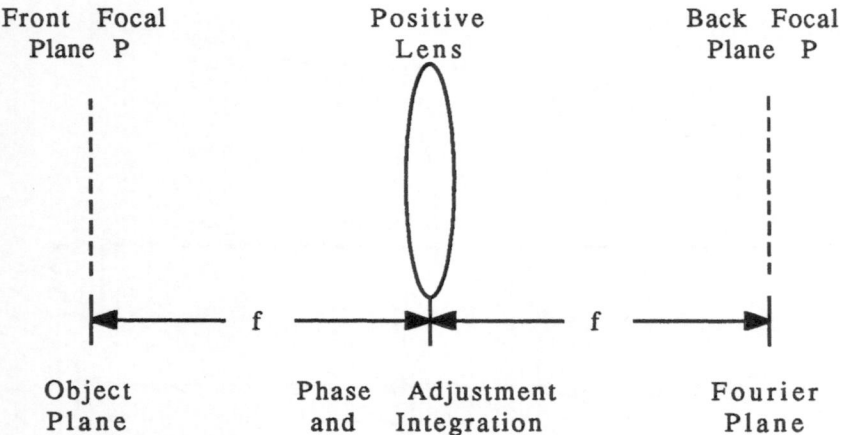

Figure 3.2: Optical Fourier transform system.

Consider, for example, a unit amplitude plane wave propagating along the z-axis incident on a rectangular aperture which represents the transparency. The plane wave can be expressed analytically as $\phi(x,y) = e^{j2\pi z/\lambda}$ and the transmission function of the rectangular aperture as

$$t(x,y) = \begin{cases} 1, & \text{if } -L \le x \le L \text{ and } -W \le y \le W; \\ 0, & \text{else.} \end{cases} \qquad (3.3)$$

The light distribution just past the rectangular mask is given by the product $\psi = t\phi$. The light distribution in the back focal plane of the Fourier transform system is given by the convolution in frequency $\Psi = T \otimes \Phi$. Detection of the intensity in the Fourier plane will produce the magnitude squared of the light distribution. Because the illumination ψ is a plane wave which is a constant with respect to x and y, the Fourier transform Ψ is a delta function. This implies that plane wave illumination will modulate the transmission function of the transparency onto the light distribution within a constant phase factor. Figure 3.3 displays the shape of this function which can also be expressed analytically as

$$\Psi(x,y) = -4ALW \text{ sinc} \left(\frac{2\pi x L}{\lambda f} \right) \text{ sinc} \left(\frac{2\pi y W}{\lambda f} \right) \qquad (3.4)$$

where sinc(t) equals sin(t)/t. Note, by decomposing the two-dimensional rectangle function $t(x,y)$ into the product of two one-dimensional rectangle

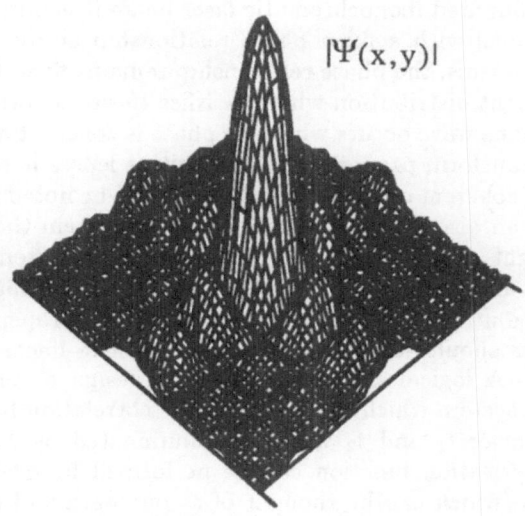

$|\Psi(x,y)|$

Figure 3.3: Fourier transform of a rectangular apreture.

functions, the required integration can be accomplished as a series of one-dimensional Fourier transformations.

Placement of the object transparency in the front focal plane creates an exact Fourier transform relation with the light distribution in the back focal plane. When doing theoretical analysis, this relation simplifies the mathematics. In practice, placing the object transparency closer to the lens (or even between the back focal plane and the lens) reduces the effects from the finite extent of the lens. Limitations caused by the physical size of the lens system are known as vignetting. Moving the object out of the front focal plane introduces an additional phase factor in the Fourier relation which can usually be ignored because the detected intensity (magnitude squared) of the light distribution is not changed by a phase modification. For object transparencies placed between the lens and the back focal plane, a similar phase modification term appears as well as a scaling of the coordinate system in the back focal plane. This scaling allows the size of the light distribution in the back focal plane to be controlled. The back focal plane is said to contain the "Optical Transform" of an object transparency which is modulating the beam but is out of the front focal plane. The optical transform is related to the Fourier transform by an appropriate modification of the phase and scale. To simplify our analysis, optical systems will be constructed to perform exact Fourier transforms.

Using a collimated monochromatic laser beam it is possible to generate light distributions with a fixed phase relationship across the beam. For stable lasing systems, the phase relationship remains fixed for a reasonably long time. A light distribution which satisfies these conditions is known as coherent. A plane wave occurs when the phase is constant across the beam. The Fourier transform property of thin positive lenses is the fundamental relationship of coherent optical analysis. It should be noted that light which is incoherent can also be described using linear system theory. Instead of linearity in light distribution, however, incoherent systems are linear in intensity. Our discussions will be restricted to coherent optical processing.

Appropriate utilization of the Fourier transform properties of coherent optical systems should allow the implementation of linear signal processing functions. A logical starting point is to design a generic linear signal processing system which can compute the correlation function between two transparencies t_1 and t_2 which are illuminated by a coherent plane wave. The correlation function can be performed by overlaying the two transparencies (which can be thought of as one data and one filter transparency) in the front focal plane of a Fourier transform lens system. In the back focal plane appears Ψ, the Fourier transform of the product of the two transparencies. Therefore, at the origin in the frequency domain is the correlation for this particular position. That is,

$$\Psi(0,0) = \int\int t_1(x,y)\, t_2(x,y)\, e^{-j\frac{2\pi}{\lambda f}(x\cdot 0 + y\cdot 0)}\, dx\, dy$$

$$= \int\int t_1(x,y)\, t_2(x,y)\, dx\, dy \tag{3.5}$$

By translating the data film across the aperture of the reference (or vice versa) a correlation is performed for every point. The Fourier transform is being used to compute the integral for the correlation operation. Figure 3.4 shows one correlation processing geometry which uses coherent optical techniques. The convolution of two transmission functions can be implemented by flipping one of the transparencies and then performing the scanning correlation operation.

Optical correlation would be a much more attractive computational technique if the physical scanning of the transparencies could be eliminated. By adding a second Fourier transform lens placed a distance f from the back focal plane of the first lens, it is possible perform the inverse Fourier transform. Consequently, frequency domain filters (transparencies in the frequency plane) can be designed for a coherent optical processing

Data   Fourier   Correlation
Input   Transform   Output at
Plane   Lens   Origin

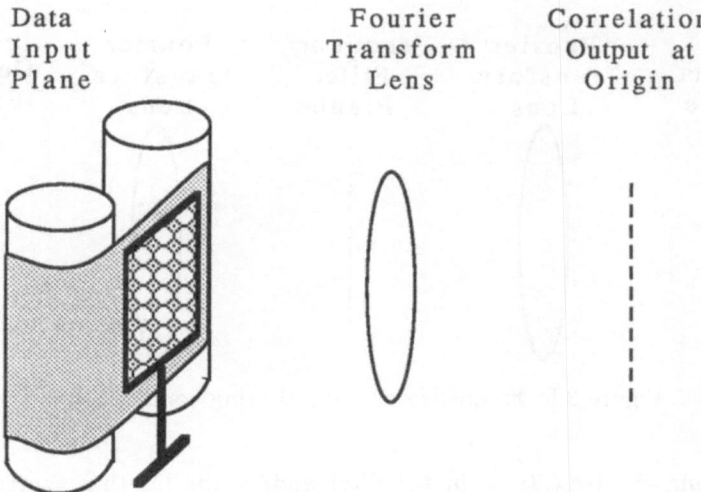

Figure 3.4: Correlation implementation using scanning.

system. A possible configuration for frequency domain filtering is shown in Figure 3.5. Because these are linear filters they are equivalent to some correlation implementation but do not require scanning the transmission plates. In addition, the frequency interpretation often provides a more intuitive description of the filtering procedure.

There are numerous applications which have benefited from using two-dimensional frequency domain filters. Examples of these filters including low pass, high pass, and knife edge filters are given in Figure 3.6. Just as with linear digital filters it is often instructive to calculate the impulse (point) response associated with the frequency domain description. The point response corresponds to the inverse Fourier transform of the frequency domain template.

Because it is convenient to fabricate binary (block/clear) filters, there are usually sharp discontinuities in the frequency definition of the filter. These sharp transitions and the clipping of the light beam by the limiting aperture of the lens system cause side-lobes in the point response. The spreading of energy over an area will produce an annoying artifact in the final image which is called ringing. These effects are completely analogous to the ringing produced when digital filters are implemented. The height of the side-lobes can be reduced relative to the main-lobe amplitude by

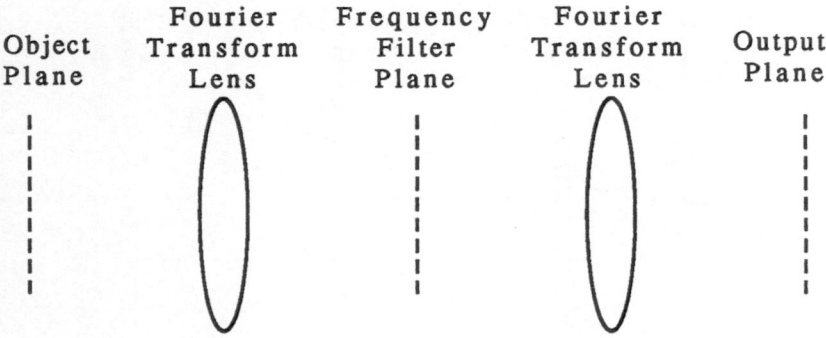

Figure 3.5: Frequency domain filtering configuration.

smoothing the transitions in the filter and/or the limiting aperture. In digital processing this is known as window weighting or simply windowing. In optics this is called apodization, which is derived from the greek language and means "remove the foot". Because this smoothing operation not only reduces side-lobes but also increases the width of the main-lobe, the trade-off between ringing and resolution must be considered in window selection.

In summary, film transparencies are used to store both the data and the filter, coherent illumination modulates the system, and the frequency domain is used to implement the two-dimensional linear filter operation. Adapting these standard techniques to the specific problem of processing SAR data is the topic of the next section.

3.2   SAR Processor

The use of film for input and output in optical systems provides a convenient and efficient mass storage medium for the raw and processed SAR data. Recall that the Seasat experiment collected 13,680 range samples per echo at a pulse repetition frequency of 1647 Hz. This represents more than 22 million samples a second. Fortunately, the data can be transferred directly to film from the receiver hardware. A frequency shifted version of the received radar echo signal (with a bias) is used to intensity modulate a crt which exposes film on a synchronized scanner. This type of data recording system is shown in Figure 3.7. Coherent illumination of the resulting transparency will modulate the information onto the light beam for processing.

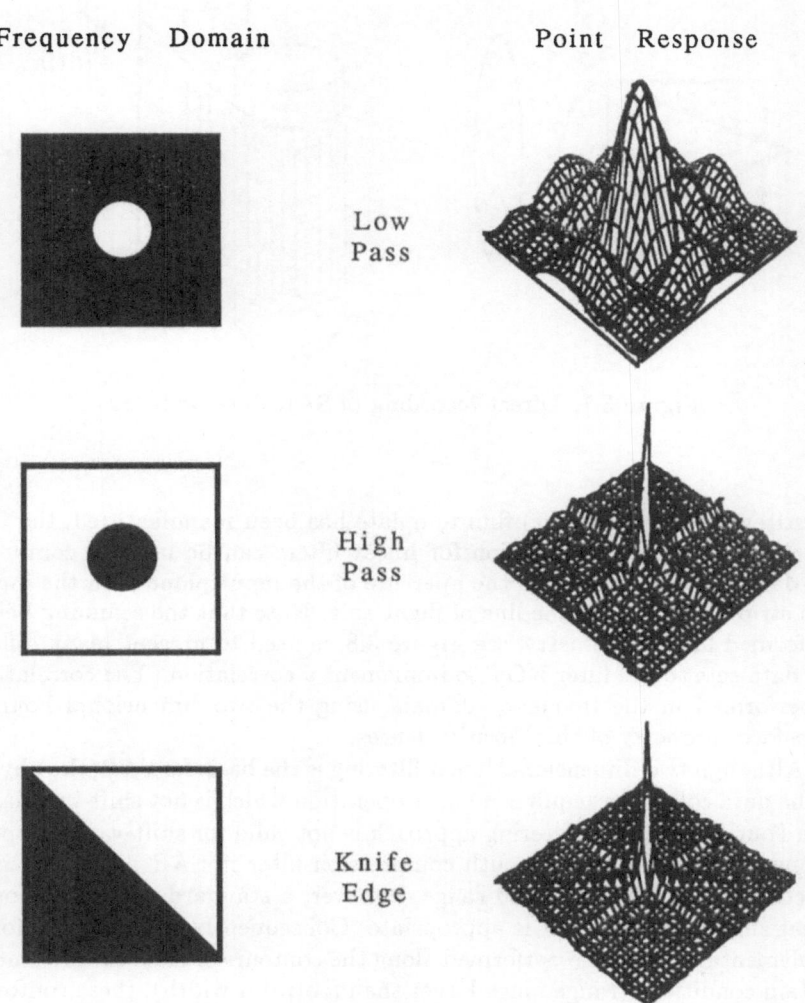

Figure 3.6: Frequency domain filters and associated point response.

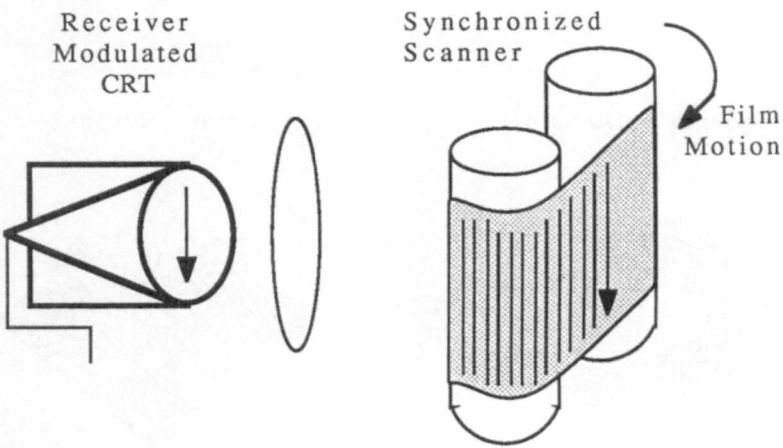

Figure 3.7: Direct recording of SAR data on film.

After a reference (filter) film template has been manufactured, the frequency domain implementation for linear filters can be used to compress the data as it scans through the aperture of the input plane with the input film strip motion along the line of flight axis. Note that the scanning being performed in this geometry, see Figure 3.8, is used to present many different data sets to the filter NOT to implement a correlation. The correlation is performed in the frequency domain using the two-dimensional Fourier transform property of thin positive lenses.

Although two-dimensional linear filtering is the basis for SAR, the physics of the data collection requires a linear operation which is not shift-invariant. The Fourier transform filtering approach is not valid for shift-varying operations. Specifically, the azimuth compression filter in SAR processing is a function of range. For a fixed range, however, a standard one-dimensional linear shift-invariant filter is appropriate. Consequently, Fourier transform implementations can be performed along the contours of fixed range. Under certain conditions (range much larger than footprint width), these contours of constant range are approximately lines on the data film parallel to the direction of flight or azimuth axis. This restriction on the range migration can be relaxed considerably by tilting certain components in the optical system. These complications, as well as the radar echo pulse compression (matched filter in the range direction), will not be incorporated in our ini-

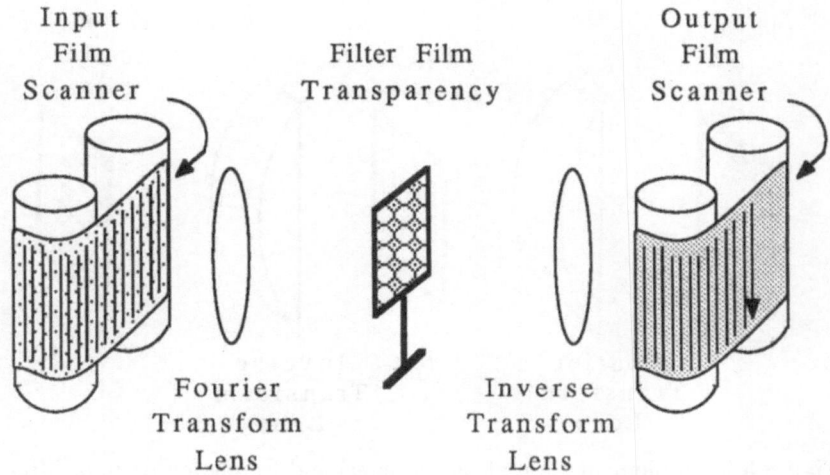

Figure 3.8: Frequency domain filter with scanning input.

tial system so that the fundamentals of the azimuth compression technique can be emphasized.

A one-dimensional Fourier transforming lens can be fabricated by replicating a radially symmetric slice from a two-dimensional Fourier transforming lens along a line instead of rotating it about the optic axis. Because of its basic shape, this lens is often called a cylindrical lens. A geometry for performing a collection of one-dimensional frequency domain filters in parallel is shown in Figure 3.9. Note that the film transmission function for the filter reference is oriented in the same direction (vertical) as the transforms performed by the cylindrical lens. dimension.

In SAR, the effective operation of the reference film transparency is to provide the matched filter required to compress the azimuthal signal. Because the along track signal is essentially modulated linearly in frequency, the compression filter implements a quadratic phase shift. A phase shift can be implemented using glass to provide the appropriate delay—a parabolic shaped piece will implement a quadratic function. The range dependent nature of the coefficients of the quadratic phase delay require varying the curvature of the paraboloid along the range dimension—a conical slice of glass is appropriate. The collection of one-dimensional filters performed using this conical lens represents a correlation type geometry for the SAR processor. As shown in Figure 3.10, the one-dimensional Fourier transform lens (cylindrical) is used to perform the integral across the quadratic

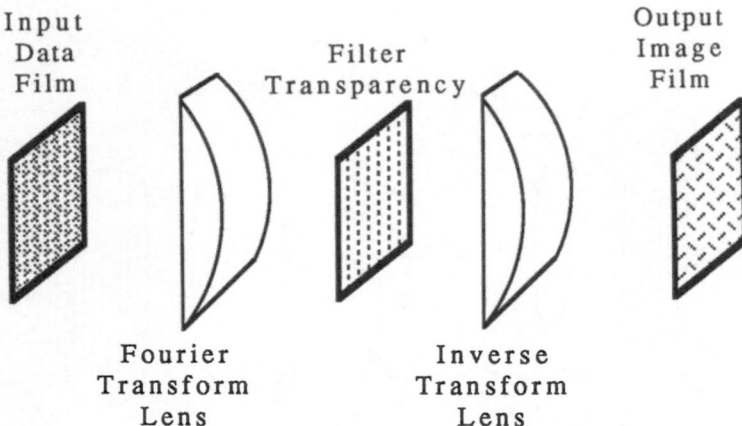

Figure 3.9: Implementation of one-dimensional frequency domain filters.

phase delay aperture for the correlation. The output for the bank of range dependent filters appears along a line through the origin of the frequency dimension.

The system in Figure 3.10 appears to be complete, however, there are several details which need to be addressed. Specifically, the component of the data in the range dimension (vertical) is not imaged (focused) onto the output film. This is similar to using a slide projector without a lens. The geometry in Figure 3.11 is proposed to incorporate range imaging without changing the functionality of the system. The cylindrical lens has been rotated one quadrant from the previous geometry and a spherical lens has been added. If each of these lenses is treated as performing a Fourier transform, then the combined effect remains a one-dimensional Fourier transform in the azimuth dimension. This is possible because the spherical lens inverse transforms the range transform performed by the cylindrical lens and forward transforms the azimuthal dimension. Treated as an imaging system, the conical lens focuses the azimuthal image to infinity, the cylindrical lens focuses the range data to infinity, and the spherical lens brings both dimensions back to focus at the output plane where the film motion is synchronized with the input.

There are many similarities between this optical processing system and electronic beam forming. The conical lens implements the appropriate range dependent phase adjustment, the Fourier transform lenses perform the coherent summation, and the output film intensity detects (recall that

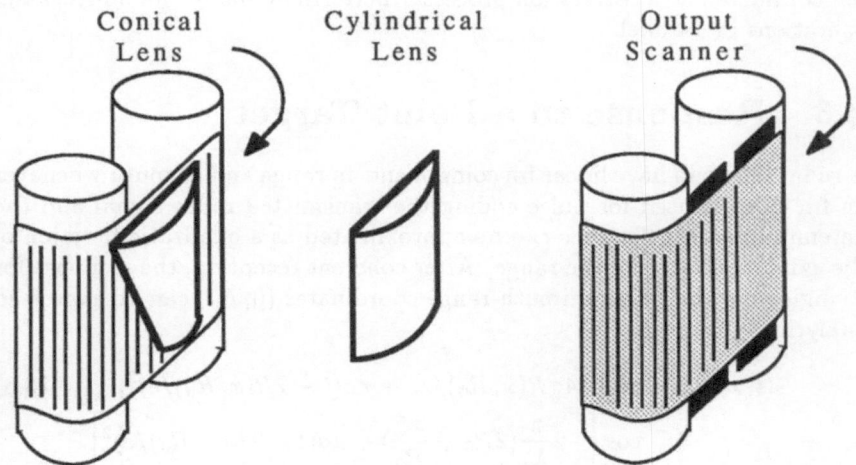

Figure 3.10: Optical SAR processor using a conical lens.

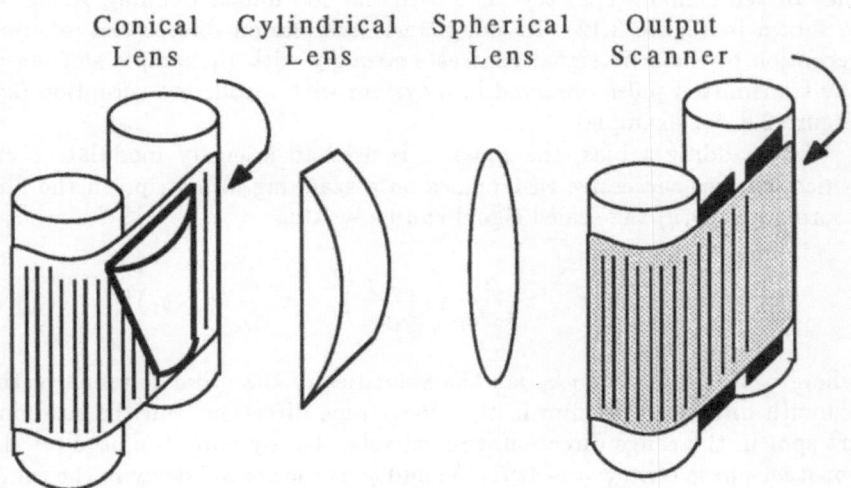

Figure 3.11: Optical SAR processor.

intensity is the magnitude squared of the wave amplitude) the final image. Because the data film scans across the input aperture, the system is often considered as a correlation processor performing many one-dimensional operations in parallel.

3.3   Response to a Point Target

A radar echo will have linear fm components in range and azimuth whenever an fm chirp is used for pulse coding the transmitted radar signal and the antenna to target distance can be approximated as a quadratic function of the azimuth offset and the range. After coherent reception, the response for a single point target at azimuth-range coordinates $(0, R_0)$ can be expressed analytically as

$$
\begin{aligned}
s(t, x) &= \cos[-4\pi R(x, R_0)/\lambda_0 + \pi a(t - 2R(x, R_0)/c)^2] \qquad (3.6) \\
&\approx \cos\left[-2\frac{\pi}{\lambda_0}(2R_0 + \frac{x^2}{R_0}) + \pi a(t - 2R(x, R_0)/c)^2\right],
\end{aligned}
$$

where t is the round trip echo delay, x is the azimuthal position, $R(x, R_0)$ is the slant range $\sqrt{R_0^2 + x^2}$, c is the speed of light, λ_0 is the carrier wavelength of the radar, and a is the transmitted chirp rate. Recall that $R(x, R_0)$ is usually approximated by $R_0 + .5\, x^2/R_0$ for R_0 much greater than x. A plot of the signal $s(t, x)$ together with the coordinate defining geometry is shown in Figure 3.12. Te along-track modulation due to the coherent reception of the echo signal contrasts strongly with the simple shifting of the transmitted pulse observed in a system with incoherent reception (see Figure 2.8, for example).

After adding a bias, the signal s is used to intensity modulate a crt which exposes successive raster lines on a scanning film strip. In the film coordinates (p, q) the scaled signal can be written

$$
f(p, q) = b + \cos\left[\psi - \pi\frac{2}{\lambda_0 R_0}\left(\frac{v_{ant}\, p}{v_{film}}\right)^2 + \pi\frac{a}{v_{crt}^2}(q - q_\tau)^2\right], \qquad (3.7)
$$

where v_{ant}, v_{film}, and v_{crt} are the velocities of the radar antenna in the azimuth direction, the film in the along-track direction, and the scanning crt spot in the range direction, respectively. The symbol ψ is used for the constant phase term $\psi = -4\pi R_0/\lambda_0$ and q_τ is the spatial delay in the range direction for the point target. The assumption that q_τ is independent of p and q is justified by the significant difference between the velocities of the spacecraft and the propagating signal. That is, the satellite does not move a

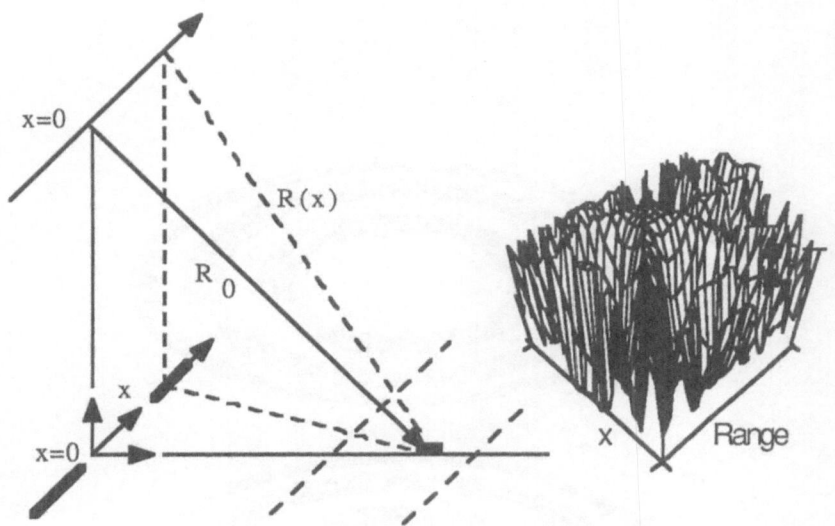

Figure 3.12: Linear fm chirp response to a point target.

significant distance during reception of a single radar echo. A hard-limited (clipped) version of the data film for the three cases: a) negligible range curvature; b) range curvature; and c) both range curvature and walk is given in Figure 3.13. The severe bending (curvature) of the point response is characteristic of spaceborne radar systems but is usually negligible in aircraft systems. As was described in Chapter 2, compensation for the linear migration of the antenna position (range walk) is also included in most spaceborne systems. In order to simplify the optical system, we will consider the response for negligible range curvature and no range walk.

When illuminated with coherent light, the point response with negligible range curvature will behave like a Fresnel lens (zone plate) and focus the incident light. Consider as a simple case the circularly symmetric Fresnel zone plate of Figure 3.14 with transmission function $b + a \cos[\pi\alpha\rho^2]$. An incident plane wave $e^{j\frac{2\pi}{\lambda_i}z}$ of wavelength λ_i traveling along the z-axis is transformed to

$$b\, e^{j\frac{2\pi}{\lambda_i}z} + .5\, a\, e^{j\frac{2\pi}{\lambda_i}z}\, e^{\pi\alpha\rho^2} + .5\, a\, e^{j\frac{2\pi}{\lambda_i}z}\, e^{-\pi\alpha\rho^2} \qquad (3.8)$$

after the zone plate. These three terms represent an attenuated plane wave propagating along the z-axis (zero-order or DC component), a diverging spherical wave (producing a virtual image), and a converging spherical wave

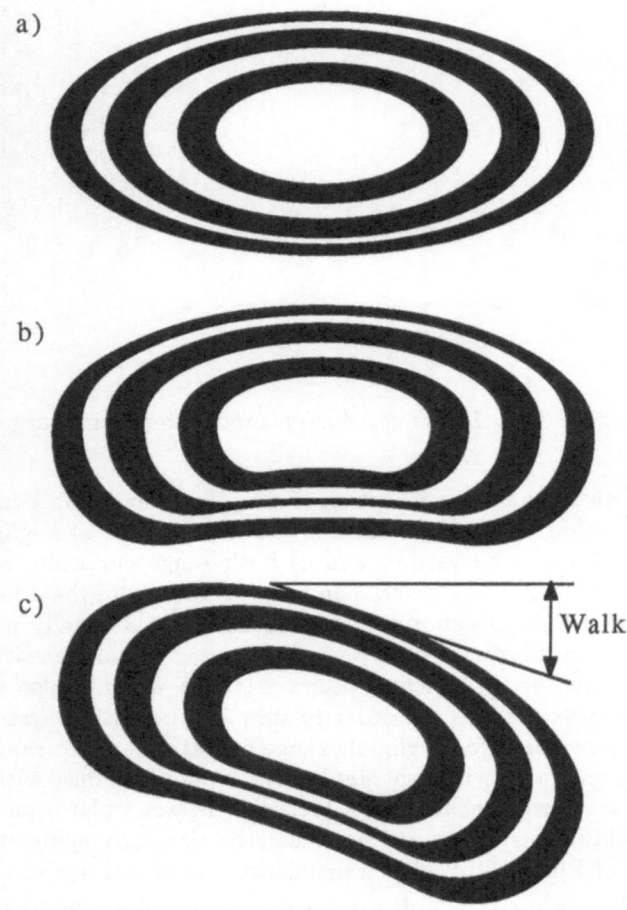

Figure 3.13: Zone plate representation for the point response with a) negligible range curvature; b) range curvature; and c) both range curvature and walk.

Figure 3.14: Circularly symmetric Fresnel lens.

(producing a real image). Comparison of the last two terms in Equation 3.8 with the expression

$$e^{-j\frac{\pi}{\lambda f}\rho^2} \tag{3.9}$$

for a spherical wave with focus f as it leaves a thin lens implies that this zone plate is represented by two lenses with focal points at $\pm(\lambda_i \alpha)^{-1}$. The two spherical wavefronts are sometimes called the plus and minus one orders.

The concept of lens-like focusing power for a circularly symmetric Fresnel plate can be extended to rectangular format zone plates. An incident plane wave can be focused with different powers along each coordinate axis producing two different focal planes. The transmission function $b + a \cos[\pi(\alpha u^2 + \beta v^2)]$ converts an incident plane wave to an attenuated plane wave, a diverging wave, and a converging wave with different focal lengths in each dimension. This can be expressed as

$$be^{j\frac{2\pi}{\lambda_i}z} + .5\, a\, e^{j\frac{2\pi}{\lambda_i}z}\, e^{\alpha u^2 + \beta v^2} + .5\, a\, e^{j\frac{2\pi}{\lambda_i}z}\, e^{-(\alpha u^2 + \beta v^2)} \tag{3.10}$$

and is displayed in Figure 3.15.

In a SAR system, linear frequency modulation of the transmitted pulse and quadratic phase delays along the azimuth axis produce a data transparency for a point target which is equivalent to a rectangular zone plate. The focusing power of the SAR data film in the range and azimuth dimen-

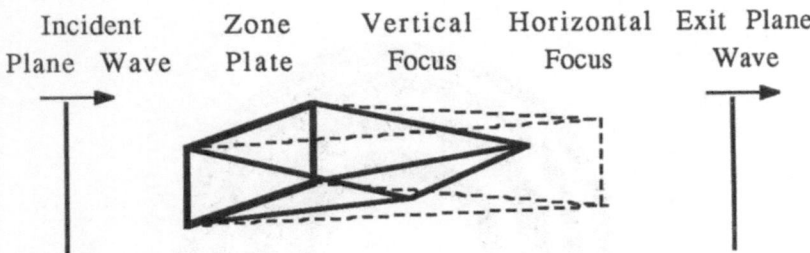

Figure 3.15: Focal properties of a rectangular zone plate.

sions are given by

$$f_{range} = \frac{v_{crt}^2}{\lambda_i a} \quad \text{and} \quad f_{azimuth} = \frac{R_0\, v_{film}^2 \lambda_0}{2\lambda_i\, v_{ant}} \qquad (3.11)$$

Because the focal power in the azimuthal dimension changes as a function of range, compensating optical components as in Figure 3.16 are required to image the waves produced by the transparency onto a plane. The conical lens has a focal length equal and opposite the azimuthal focus of the zone plate and provides a means of transporting this component of the signal to infinity. The cylindrical lens performs a similar operation by imaging the range signals to infinity. Finally, the spherical lens collects the data in both dimensions and reimages it at its focal plane.

Note that a conical lens is required because the focal power of the transparency in the azimuth dimension varies as a function of the range dimension. The range dependence of the linear chirp rate of the azimuth modulation of the zone plate results in a tilted focal plane. This implies that the conical lens could be replaced by a tilted cylindrical lens imaging the azimuth focal plane to infinity (both are "matched" to the pulse being compressed). The range signal provides essentially uniform focal power over the transparency and therefore directs the incident light onto a plane. The tilted cylindrical lens compression technique is illustrated in Figure 3.17. Because cylindrical lenses are simpler to fabricate than conical lenses, this geometry is very cost effective. In addition, different tilts can be implemented using this scheme that might have required fabrication of an entirely new conical lens using the previous geometry.

It is interesting to compare the physical functions of this reconstruction geometry with the actual wavefronts encountered in the radar system. The tilt of the cylindrical azimuth lens is analogous to the perspective view of the ground patch illuminated by the radar. The curvature of this lens

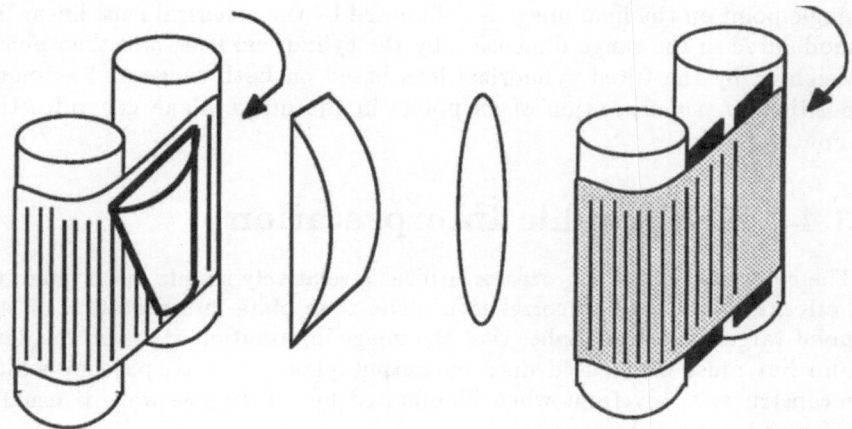

Figure 3.16: Simple optical SAR processor.

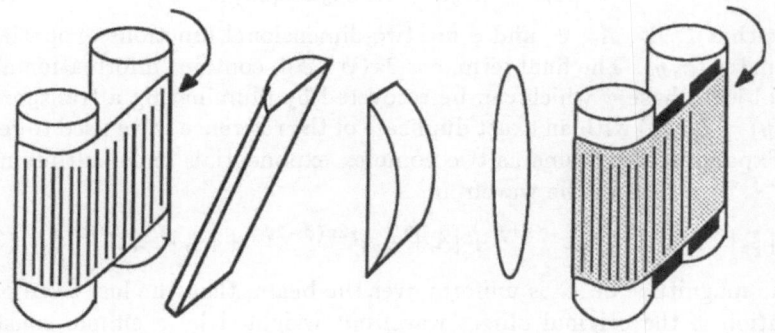

Figure 3.17: SAR processor with tilted cylindrical lens.

compensates for the difference in phase between the spherical wave reflected from a target and the line of flight reference. A similar compensation is required for range pulse compression. By reversing the illumination in the optical system it is possible to trace the generation of the data film. A single point on the final image is collimated by the spherical lens, linear fm modulated in the range dimension by the cylindrical lens, and then phase weighted by the tilted cylindrical lens based on both range and azimuth position. A superposition of all points in the image plane generates the input data film.

3.4   Holographic Interpretation

The conventional SAR processor utilizes a relatively simple lens system to perform a shift-varying correlation. The zone plate interpretation of the point target response implies that the image information exists on the data film but must be focused onto an output plane. A transparency which reconstructs a wavefront when illuminated by a reference wave is usually referred to as a hologram.

The conventional geometry for recording optical holograms, Figure 3.18, utilizes a collimated plane wave as the illumination and reference wavefront. The object reflection is mixed with the reference to allow the recording of phase information about the wavefront using the intensity detecting property of film. If the reflected object wave is $A_o\, e^{-j2\pi\phi}$ and the reference wave is $A_r\, e^{-j2\pi\psi}$, then the film records the intensity of the sum of the two waves or

$$I = |A_r|^2 + |A_o|^2 + 2A_r A_o \cos[2\pi(\psi - \phi)]. \tag{3.12}$$

Note that I, A_r, A_o, ψ, and ϕ are two-dimensional functions of spatial coordinates (x, y). The final term, $\cos[2\pi(\psi - \phi)]$, contains information about the object phase ϕ which can be recovered by illuminating a transparency $t(x, y) = I(x, y)$ with an exact duplicate of the reference wave used to record I. Expanding the cosine as two complex exponentials and multiplying by $A_r e^{-j2\pi\psi}$ results in the wavefront

$$\left\{ |A_r|^2 + |A_o|^2 \right\} A_r e^{-j2\pi\psi} + |A_r|^2 A_o e^{j2\pi(\phi - 2\psi)} + |A_r|^2 A_o e^{-j2\pi\phi}. \tag{3.13}$$

If the magnitude of A_r is uniform over the beam, then the last term of the equation is the original object wavefront weighted by a simple constant. Under proper conditions the wavefronts represented by the other terms in this equation will be spatially separated from the reconstructed image.

In a SAR system, the stable local oscillator (STALO) on board the aircraft provides a reference function which allows recording the phase information of the return echo. The STALO performs the same function for a

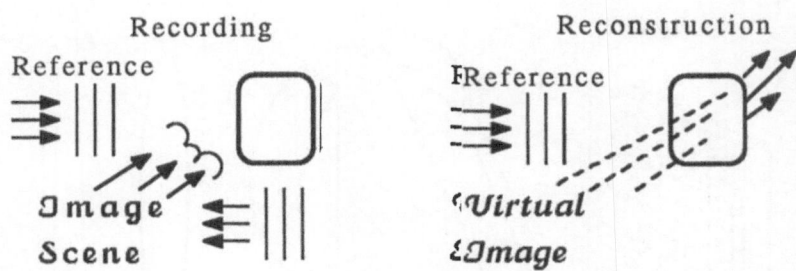

Figure 3.18: Geometry for holographic recording and reconstruction.

SAR system that plane wave laser references perform when making optical holograms. As was noted when discussing zone plates, the optical SAR processor reconstructs the imagery by providing a duplicate of the reference signal (scaled to the dimensions of the film plane). This reconstruction can be interpreted as a correlation using information processing techniques, a zone plate (Fresnel type) lens, or a hologram. Because holograms more closely associated with wavefront reconstruction, SAR processors are placed in this category even though the other two interpretations are equally valid.

3.5 Advances in Optical Processing

Although optical systems perform processing at real time rates, the use of film often reduces the effective image reconstruction time. The film scanners required by the conventional optical SAR processor are large and heavy, making airborne systems expensive. Reducing these negative aspects of optical processing is a heavily pursued area of research which has been relatively successful. In this section two systems are described which replace the input film strip, one with a two-dimensional spatial light modulator followed by a conventional correlation processor is described and the other with a one-dimensional acousto-optic input cell followed by a spatial range and temporal azimuth correlation.

There are several commercially available spatial light modulators (SLM) which provide the ability to modulate a two-dimensional pattern onto a laser beam. The speed and quality of these devices is rapidly improving as liquid crystals and other related technologies are better understood. SAR data stored on computer disk is transferred to the drive electronics of the SLM which modulates the light beam and is then processed by a

Figure 3.19: SAR processor using input from a spatial light modulator.

conventional optical correlator. A sensor array placed in the output plane converts the optical signal to an electronic signal for storage and additional processing by digital computer. The basic configuration for this type of system is shown in Figure 3.19. Recall that the importance of this processing approach is the elimination of photographic film from the system. Current limitations include the speed for writing signals onto an SLM and the associated decay (storage) time after the signal has been written.

Another type of SAR processor utilizes a one-dimensional acousto-optic (A-O) cell for data input and a two-dimensional charge coupled device (CCD) array for azimuth signal accumulation and output. As diagrammed in Figure 3.20, after a range echo is propagated down the A-O cell it is "flash" illuminated by a laser diode and then collimated for conventional pulse compression by diffraction. The light which is imaged onto the array is constant along the vertical (azimuthal) axis and can be weighted as a function of this position to provide focus along the synthetic aperture. The azimuth correlation is accomplished by shifting the rows (horizontal elements) down after each range "flash" and then adding the results of the next exposure to the stored value. The final row of the CCD array is a compressed range strip of the SAR image. Range migration, multi-look processing, and Doppler offsets can be corrected by making straight forward adaptations of this geometry.

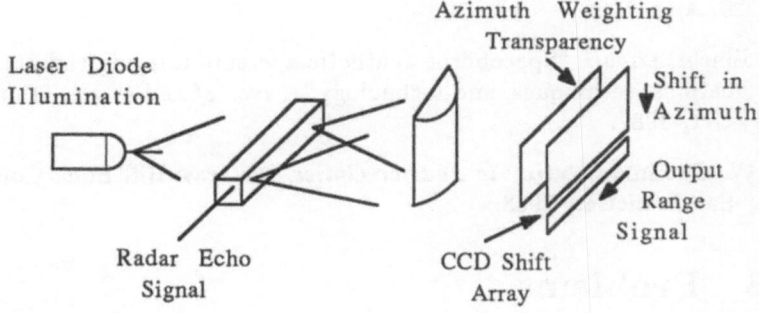

Figure 3.20: SAR processor using acousto-optic input cell.

3.6 Summary of Optical Processing Techniques

Conventional SAR processors exploit the benefits of high speed parallel processing offered by optical systems. The information processing algorithm can be interpreted as a shift-varying correlation, holographic reconstruction, or focusing fresnel zone plate lens. Historically, optical interpretations and optical processors have made SAR an understandable and workable technology. Restrictions imposed by using film in these systems can be overcome with more exotic optical components such as spatial light modulators and acousto-optic cells. Before exchanging your digital array processor for an optics bench with a Fourier transforming lens, consider that transferring information from digital storage to an optical medium is expensive in dollars and in time. Even when the processing can be performed at the speed of light, the optical to digital conversion may be extremely slow and less precise than current digital floating point representations. As the capability to interface between general purpose digital computers and special purpose hardware (optical, digital, or analog) improves, new avenues for SAR reconstruction and post-processing become available.

3.7 References

W. M. Brown and L. J. Porcello, "An introduction to synthetic-aperture radar," *IEEE Spectrum*, pp. 52–62, Sept. 1969.

L. J. Cutrona, *et. al.*, "On the application of coherent optical processing techniques to synthetic-aperture radar," *Proc. of IEEE*, vol. 54, no. 8, Aug. 1966

C. Elachi, *et. al.*, "Spaceborne synthetic-aperture imaging radars: applications, techniques, and technology," *Proc. of IEEE*, vol. 70, no. 10, Oct. 1982.

J. W. Goodman, *Intro. to Fourier Optics*, McGraw-Hill Book Company, San Francisco, 1968.

3.8 Problems

P3.1 When using a single positive lens, where will the image of an object located at infinity appear?

P3.2 Suppose that the one-dimensional functions a and b have Fourier transforms A and B, respectively. What is the two-dimensional Fourier transform (optical transform) of the function $c(x,y) = a(x)b(y)$?

P3.3 Suppose that $S(u,v)$ is the Fourier transform of $s(x,y)$. What is the Fourier transform of   a) s translated by (x_0, y_0),   b) s scaled to $s(ax, by)$, and   c) s rotated an angle θ.

P3.4 If $s(x,y)$ is restricted to be a real-valued function What properties does the Fourier transform $S(u,v)$ have?

P3.5 What effect does a frequency domain filter (two-dimensional) which blocks half the Fourier space (through the origin)? Hint: think of Hilbert transformations or I-Q reception.

P3.6 Suppose a photo is compiled from a collection of horizontal film strips (these mosaic images are typical of telemetered scenes from early satellite systems). Design a frequency domain filter to remove the horizontal striping. Perhaps the problem is more easily described as removing a regular pattern of raster scan blanks from an image.

P3.7 Why is a system which uses a frequency domain filter for blocking the DC (zero frequency) component of a phase-only input transparency called a dark ground system?

Chapter 4

Related Algorithms: An Overview

Synthetic aperture radar processing can be interpreted using a combination of concepts from many disciplines including radar (pulse echo, Doppler, and strip map systems), optics (wavefront reconstruction, zone plate lenses, and holography), communication theory (sampling criteria and coherent reception), antenna theory (resolution, propagation, and delay-add), and linear system theory (convolution, Fourier techniques, and shift-variant operations). It is not surprising that this technological arsenal has also been aimed at applications which are similar to SAR. Several of these applications including Synthetic Aperture Focusing Techniques in Ultrasonic Testing (SAFT-UT), tomography, spotlight mode SAR, and inverse SAR (imaging rotating targets with a fixed antenna), are briefly described in the sections which follow. Development of an intuition for the physics of the application will be emphasized not the data processing. Hopefully future research will unify the analysis and processing required for this class of problems so that the mathematics and computer architectures will be appropriate for several of these important applications.

4.1   Ultrasonic Inspection—SAFT-UT

There are some fundamental differences between microwave (radar) and ultrasonic (acoustic) wave propagation which influences the associated imaging techniques. Radar frequencies are typically several orders of magnitude greater than ultrasonic frequenies. For radar systems, the high frequencies require special detection and recording techniques to maintain phase coherence during data collection; ultrasonic systems in the 1 to 10 MHz range often directly record the propagating disturbance using high-speed analog to digital converters (ADCs). Radar imaging is usually performed in a relatively uniform medium with a single index of refraction and a single mode of propagation while ultrasonic imaging uses media with variable indices of refraction. This fact leads to a number of complicating phenomena such as beam steering caused by refraction, multiple-mode propagation caused by mode conversion, and guided-wave phenomena caused by material anisotropy. Radar has its own set of analogous problems but seems to have overcome these obstacles to the degree that synthetic aperture imaging can be achieved without addressing each of the problems individually in the processing—ultrasonic imaging has not been as uniformly fortunate although it has been applied successfully to many problems. This section presents a brief discussion of synthetic aperture focusing with ultrasound.

When describing ultrasonic phenomena it is usually assumed that the material to be inspected is isotropic (the average acoustic properties can be characterized as a scalar quantity—does not vary with the direction of propagation) and homogeneous in the sense that the average acoustic properties (velocity and density) are consistent throughout the material. Variations from the average velocity and density caused by cracks and voids lead to reflected ultrasonic energy which is used to form the image. Small grain metals tend to be isotropic, however, large grain metals such as centrifugally cast stainless steel may be anisotropic.

A typical geometry for collecting ultrasonic images is shown in Figure 4.1. The transducer is incrementally positioned along a collection arc (usually a line for two-dimensional imaging or a raster for three-dimensional imaging), a pulse is transmitted, the echo is received, and stored for processing/display. The individual pulse echoes are referred to as A-scans and the juxtaposed collection of A-scans is known as a B-scan. In medical imaging applications, the sequential positioning of a single transducer is often replaced with an array of transducers for rapid sequential or simultaneous pulse echoing at many positions. The motivation for medical applications is that the "part" is often alive and therefore tends to move during inspection. Rapid data collection for parts which move unpredictably makes image reconstruction possible in these applications.

Synthetic aperture focusing refers to a process in which the focal properties of a large-aperture, focused transducer (lens or antenna), are mimicked or synthesized from a series of measurements made using a small-aperture transducer which has been scanned over a large area. The processing required to focus the data is known as beamforming, coherent summation, synthetic aperture processing or Synthetic Aperture focusing Technique for Ultrasonic Testing (SAFT-UT). The major difference between SAR and

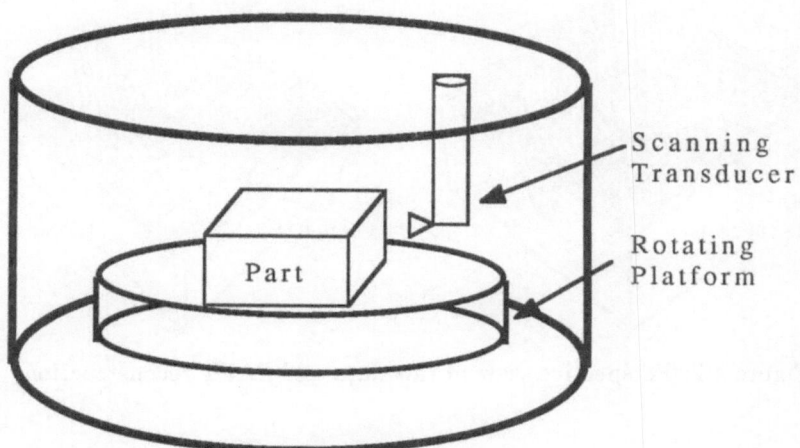

SAFT is that SAR utilizes a sophisticated azimuthal matched filter for compressing in the along-track direction while the azimuthal reference in SAFT is usually a uniformly weighted antenna pattern. The range migration correction is performed but instead of the quadratic phase (linear fm) reference common in SAR, a simple summation is performed in SAFT. The pseudo-code example for radar delay and add given in Figure 2.14 outlines the appropriate steps for processing SAFT data and the simulations shown in Figure 2.12 are valid for ultrasonic systems. Application of a simple SAFT algorithm to a data set collected using a 10 MHz ultrasonic transducer and inter-sample spacings of 26 μm in azimuth and 27.6 μm in range (50 MHz sampling frequency), is shown in Figure 4.2. The flaws are three flat bottom holes 3–4 mm wide and separated by about 0.5 mm. The SAFT algorithm enhances the data so that the three flaws are significantly easier to discriminate. The data has been subsampled to a 32 by 32 grid for resolution comparisons using perspective plots.

One of the problems with SAFT systems is the miscalculation of reflector locations due to uncompensated variations in the front surface geometry. This is analogous to the spacebourne SAR problem of modeling the platform position variations for compensation. In ultrasound systems where the transducer is oriented normal to the front surface of the part (known as normal-beam SAFT), compensating for variations in the front-surface geometry to reduce image aberrations is usually performed as part of the

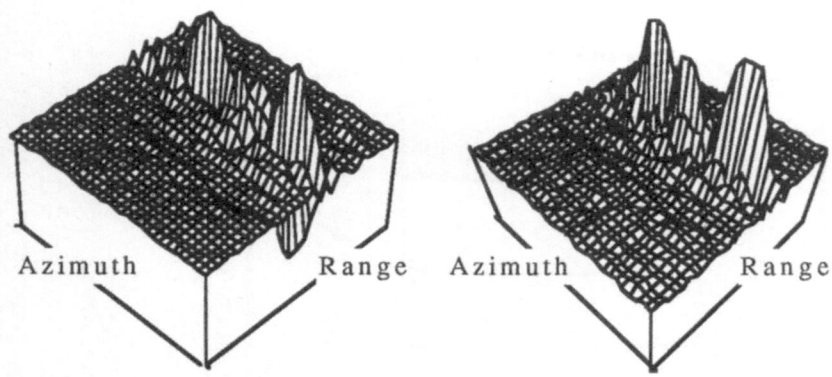

Figure 4.2: Perspective view of raw data and SAFT reconstruction.

in-line signal processing by detecting the front-surface echo, removing this contribution to the echo by subtraction (if this subtraction is done before analog to digital quantization an improvement in signal to noise will also be achieved), and shifting the echo by the appropriate delay to the front surface. The front-surface echo can also be eliminated by taking the data with the transducer oriented at an angle declined from the normal to the surface (known as angle-beam SAFT). This eliminates the large front-surface echo but introduces ambiguities caused by reflections traveling the direct path to a flaw, the double bounce path back-surface to flaw to receiver, and the triple bounce path back-surface to flaw to back-surface to receiver. Although other ambiguous paths also occur, the three mentioned paths tend to dominate in practice and an angle-beam SAFT algorithm must be able to handle all three. If accurate part geometry and orientation is available (as with numerically milled parts), then front-surface echo suppression and multi-path discrimination can both be accomplished using the a *priori* part information. When part descriptions are not available adaptive digital filtering techniques can be utilized.

In summary, the processing required for SAFT-UT reconstructions is very similar to strip-map SAR imaging. Normal-beam SAFT is analogous to the delay-sum operations described in Chapter 2 for a simple side-looking SAR system. Angle-beam SAFT corresponds to the squint-mode SAR systems which require similar processing. Ultrasonic imaging suffers from most of the short-comings common to radar imaging systems including foreshortening (the variation in apparent length of target slopes when the measurements are taken at different angles of incidence), layover (the

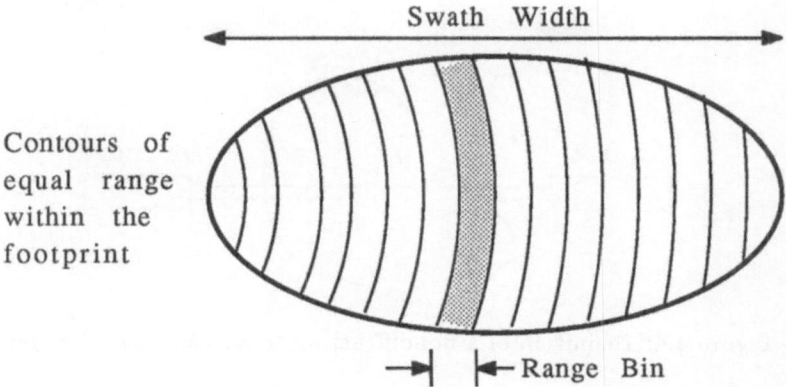

Figure 4.3: Integration of reflectivity along arcs of constant range.

reception of information from the top of a target before the echo from the bottom of the target arrives caused by slopes in the target which are greater than the radars look angle), shadows (the absence of echoed information from targets positioned behind other targets), antenna pattern weighting (modulation in the along-track direction), and sidelobes (ringing caused by the use of a finite number of Fourier coefficients to represent and process the data). Despite these limitations, high resolution SAFT imaging has been used successfully in several applications.

4.2 Tomography

The data collected by a radar system is the echo signal. The value of an echo signal at a time delay τ represents the average (integrated) reflectance of all targets in the radar illumination at a range $r = c\tau/2$, where c is the speed of light and the division by 2 accounts for the round trip propagation. If the reflectance observed by a sensor at azimuthal position x is considered as a two-dimensional function of range and ground angle $\sigma(r, \theta)$, then the received echo can be written

$$s_x(r) = \int_{beam} \sigma(r, \theta)\, d\theta. \tag{4.1}$$

The path of integration is illustrated in Figure 4.3. Estimation of the original reflectivity distribution from the echo measurements is accomplished using synthetic aperture techniques.

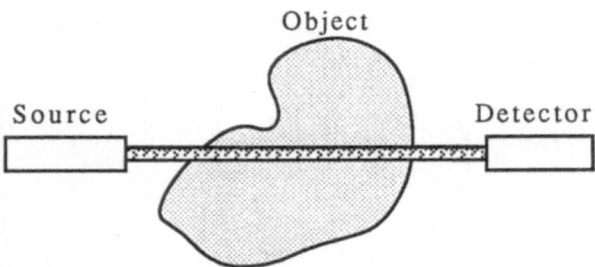

Figure 4.4: Definition of a nondiffracting tomographic projection.

Measurements known as projections are used in a tomographic imaging system to estimate a transmission distribution. A projection, which is similar to a radar echo, represents the average (integrated) transmission through a sample. Illumination of the sample is usually performed with nondiffracting sources, such as x-rays, so that the projections will be along straight lines as in Figure 4.4. Even when the inspected object is three dimensional, the tomographic reconstruction problem is often formulated as the estimation using projection data of the original transmission function in a plane. Three-dimensional transmission distributions are then synthesized by stacking two-dimensional results. In medical tomography this approach has produced incredible images of soft tissues without invasive surgery. Similarly spectacular results have been obtained in industrial non-destructive evaluation systems.

Typically a set of parallel projections are collected by physically scanning the transmitter/receiver pair in a plane or by using planar illumination together with a sensor array as in Figure 4.5 The equation describing the linear attenuation of the illumination as it propagates through the object is given by

$$p_\theta(t) \;=\; \int_{-\infty}^{\infty}\int_{-\infty}^{\infty} \sigma(x,y)\,\delta\left(x\cos\theta + y\sin\theta - t\right)dx dy. \qquad (4.2)$$

$$=\; \int_{-\infty}^{\infty} \sigma(t\cos\theta - \tau\sin\theta, t\sin\theta + \tau\cos\theta)\,d\tau$$

Note that $\sigma(x,y)$ would be referred to as the transmission function in optical applications but is usually referred to as the absorption or attenuation function for non-visible wavelengths and energy beams. For fixed t, the previous equation represents a line integral along a path which is perpendicular

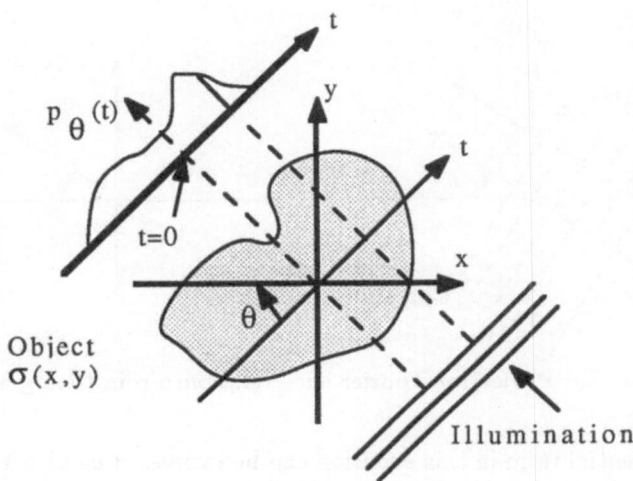

Figure 4.5: Typical geometry for a tomographic system.

to a line through the origin and rotated an angle θ from the x-axis. The argument t specifies the distance from the origin where the integration path crosses the line of the rotated x-axis. The function $p_\theta(t)$ is often called the parallel projection or Radon transform. It is useful to think of this function as a collection of point to point projections through the sample.

Recall that a tomographic system uses projection data sets to reconstruct (estimate) an image slice of the transmission attenuation function. The parallel projection $p_\theta(t)$ is important in tomography because there is an analytic relationship between its Fourier transform and the Fourier transform of the attenuation function $\sigma(x, y)$. In order to derive this important relationship, consider the Fourier transform of σ in Cartesian frequency coordinates (u, v):

$$\Sigma(u, v) = \int_{-\infty}^{\infty} \int_{-\infty}^{\infty} \sigma(x, y)\, e^{-j2\pi(ux+vy)}\, dx\, dy. \qquad (4.3)$$

If the frequency domain is represented in polar coordinates (ρ, θ) without changing the coordinate system of the attenuation function, the previous equation becomes

$$
\begin{aligned}
S(\rho, \theta) &= \Sigma(\rho \cos\theta, \rho \sin\theta) \qquad\qquad\qquad (4.4)\\
&= \int_{-\infty}^{\infty} \int_{-\infty}^{\infty} \sigma(x, y)\, e^{-j2\pi\rho(x\cos\theta + y\sin\theta)}\, dx\, dy.
\end{aligned}
$$

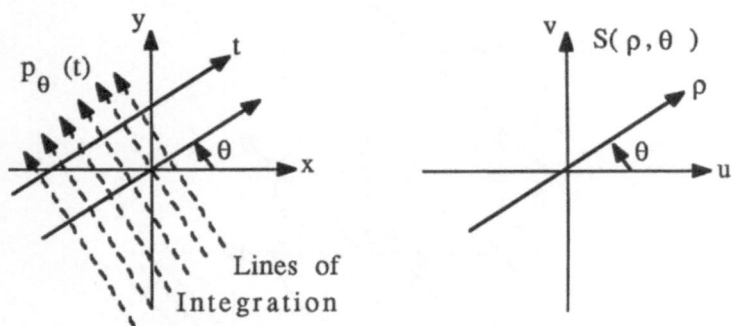

Figure 4.6: Projection-Fourier slice relationship in tomography.

The exponential term in this equation can be expressed using a Dirac delta function and an integral as

$$e^{-j2\pi\rho(x\cos\theta+y\sin\theta)} = \int_{-\infty}^{\infty} e^{-j2\pi\rho t}\, \delta(x\cos\theta + y\sin\theta - t)\, dt. \qquad (4.5)$$

Substituting this into our earlier equation and changing the order of integration results in

$$\begin{aligned}
S(\rho,\theta) &= \int_{-\infty}^{\infty} \left\{ \int\int \sigma(x,y)\, \delta(x\cos\theta + y\sin\theta - t)\, dx\, dy \right\} e^{-j2\pi\rho t}\, dt \\
&= \int_{-\infty}^{\infty} p_\theta(t)\, e^{-j2\pi\rho t}\, dt. \\
&= \Sigma(\rho\cos\theta, \rho\sin\theta). \qquad (4.6)
\end{aligned}$$

This equation relates the Fourier transform (with respect to t) of the parallel projection $p_\theta(t)$ to the Fourier transform Σ of the attenuation function $\sigma(x,y)$. In short, the Fourier transform of the parallel projection of angle θ equals the Fourier transform of the attenuation along a line through DC (frequency origin) and at an angle θ. This result is called the Fourier-slice or projection-slice theorem. Figure 4.6 illustrates the relationship between the two domains.

The data required to estimate the attenuation function is contained in a set of parallel projections taken at many different orientations. The Fourier transform of each parallel projection equals a ray through the origin of the frequency representation of the attenuation function. After the frequency space is sufficiently filled (satisfies Nyquist's Theorem), an inverse Fourier

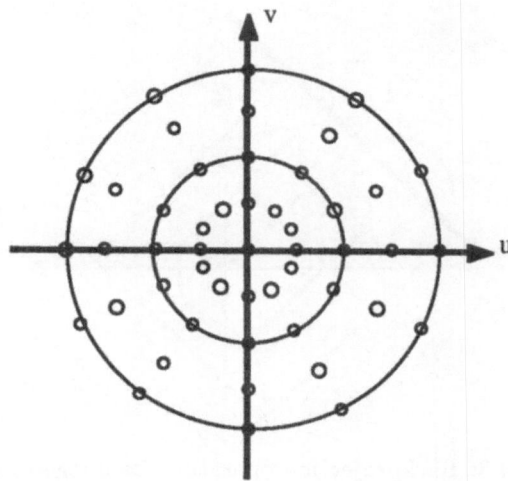

Figure 4.7: Polar grid from uniformly spaced projection samples.

transform reconstructs an estimate of the attenuation distribution. This tomographic reconstruction algorithm is called Fourier inversion.

When digital computers are used to perform Fourier inversion tomography, the projections are discrete sequences or arrays. Fast Fourier Transform (FFT) algorithms are used to transform the parallel projection $p_\theta(t)$ into a slice of the frequency domain. Uniformly spaced samples of $p_\theta(t)$ become radially uniform samples in $\Sigma(u, v)$. That is, the frequency domain samples are on a polar grid as shown in Figure 4.7.

Interpolation of the data from a polar to a rectangular grid is required before standard FFTs can be used to inverse transform the frequency representation to the spatial domain. The interpolation and FFT operations are computationally intensive and can reduce the precision of the reconstruction. Backprojection techniques are often used in place of the Fourier inversion approach to avoid interpolation in the frequency domain and reduce round-off effects. When a restricted number of projections are available, it is common to use Algebraic Reconstruction Techniques (ART) combined with some knowledge of the absorption process to estimate the attenuation distribution. Because there is a fundamental equivalence between projections and radar echoes, both of these tomographic processing strategies will be summarized as well as a matched filter interpretation of reconstruction from projections.

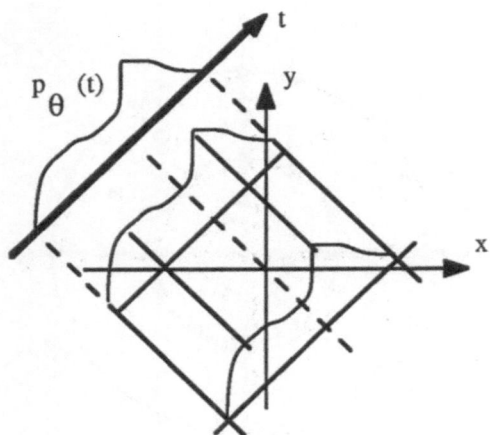

Figure 4.8: Backprojection operation for image estimation.

4.2.1   Filtered Backprojection Tomography

If only one parallel projection $p_\theta(t)$ is available to estimate the attenuation distribution, then Fourier inversion is not useful unless some property of the original image–$e.g.$, circular symmetry, is known. A "best guess" from one projection might assume that each line integral represented in $p_\theta(t)$ is the result of a uniformly distributed image. That is, set the estimate of the attenuation function equal to $p_\theta(t)$ for every position which contributed to this specific value of t. This process of "smearing" the projection back over the support of the original image is known as backprojection and is illustrated in Figure 4.8. Mathematically the backprojection estimate b_θ in polar coordinates is given by

$$b_\theta(r, \phi) = p_\theta[r \cos(\theta - \phi)]. \tag{4.7}$$

If many projections are available, the average will provide an intuitive estimate of the original image. Given a continuum of projection orientations, the backprojection summation image estimate $\hat{\sigma}_{bp}$ can be expressed as the integral

$$\hat{\sigma}_{bp}(r, \phi) = \frac{1}{\pi} \int_0^\pi b_\theta(r, \phi) \, d\theta = \frac{1}{\pi} \int_0^\pi p_\theta[r \cos(\theta - \phi)] \, d\theta. \tag{4.8}$$

Unfortunately the backprojection technique does not reconstruct the original image exactly. Because this technique is a linear procedure, determination of the correction required to eliminate reconstruction error for an impulsive input will also provide the necessary modification for arbitrary inputs. A Dirac delta function at the origin of the input image results in a summation image which is the two-dimensional impulse response

$$h_{bp}(r, \phi) = \frac{1}{\pi} \int_0^\pi \delta[r \cos(\theta - \phi)] \, d\theta. \tag{4.9}$$

To reduce this integral, we use the identity

$$\delta[f(x)] = \delta(x - x_0) \left| \frac{df}{dx} \right|_{x=x_0}^{-1}, \tag{4.10}$$

where $f(x_0) = 0$, which implies

$$h_{bp}(r, \phi) = \frac{1}{\pi |r|}. \tag{4.11}$$

The summation image produced by backprojection tomography is represented by the following two-dimensional convolution

$$\hat{\sigma}_{bp}(r, \phi) = \sigma(r, \phi) * h_{bp}(r, \phi) = \sigma(r, \phi) * \frac{1}{\pi |r|}. \tag{4.12}$$

There are various techniques for performing "filtered backprojection" which reduce or eliminate the effects of the $(\pi|r|)^{-1}$ distortion in the backprojection summation operation. Because $(\pi|r|)^{-1}$ Fourier transforms to $|\rho|^{-1}$, the correction can be done in the frequency domain by multiplication with $|\rho|$. This correction is often called rho-filtering. The circular symmetry of this filter in combination with the projection-slice theorem implies that a one-dimensional rho-filter operating on each of the projections is equivalent to the two-dimensional filter. Note that a rho-filter is neither band-limited nor absolutely integrable, implementation in the spatial domain requires an approximation usually performed by windowing the filter transfer function. Because the inherent distortion of the backprojection summation operation can be corrected with linear one-dimensional filters, filtered backprojection is a very popular algorithm used in many commercial tomographic systems.

4.2.2   Algebraic Reconstruction Techniques (ART)

For a small number of projections with a good physical model of the attenuation distribution, it may be appropriate to use an algebraic estimate of the image. The fundamental premise in algebraic reconstruction techniques is that the projections of the reconstruction should match the measured

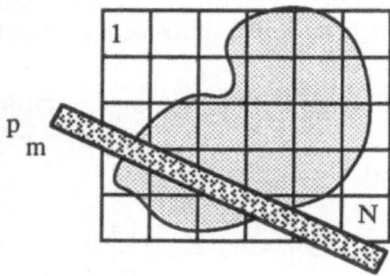

Figure 4.9: Possible geometry for algebraic reconstruction tomography.

projections. The model begins with an array $\hat{\sigma}$ of N elements to be reconstructed and a set of M projections p_m, $1 \le m \le M$. Note that discrete mathematics is used here–a deviation from the functions with continuous parameters used to derive our earlier tomography results. A potential geometry for this approach is shown in Figure 4.9. The goal is to solve for the $\hat{\sigma}_n$ such that

$$p_m = \sum_{n=1}^{N} w_{mn}\, \hat{\sigma}_n, \quad m = 1, \ldots, M. \qquad (4.13)$$

The w_{mn}'s are weights which incorporate the area of the resolution cell (pixel) and the physical width of the ray performing the line integration. Note that most of the w's are zero since only a small number of attenuation cells intersect a specific illumination ray.

ART is an iterative algorithm for solving these equations. Assume a constant for the initial guess (zero will do). Each step in the iteration incorporates data from one projection. First, determine an error that is the difference between the calculated projection and the measured projection. Second, use the error to adjust or "correct" each of the elements which contributed to that projection. Both steps are repeated until all projections are included or until convergence is achieved. The method performs best if the corrections are made one projection at a time with large angles between consecutive projections. Using vector notation to represent the m'th guess $\hat{\sigma}^{(m)}$ which equals $(\hat{\sigma}_1^{(m)}, ..., \hat{\sigma}_N^{(m)})$ and the weights $\mathbf{w}_m$ which equals $(w_{m1}, ..., w_{mN})$, then the ART is

$$\hat{\sigma}^{(m)} = \hat{\sigma}^{(m-1)} - \frac{\left(\hat{\sigma}^{(m-1)} \cdot \mathbf{w}_m - p_m\right)}{\mathbf{w}_m \cdot \mathbf{w}_m}\mathbf{w}_m. \qquad (4.14)$$

Formulating the reconstruction problem as a collection of matrix operations allows a linear algebraic analysis of the solution space. Various matrix decompositions (singular value decompositions, for example) have been used to characterize the collection of solutions (the space) which will produce the same projection meausurements. Because ART usually requires more computation time than direct methods and because the system of equations has no unique solution whenever $M < N$, iterative algebraic techniques are typically not used in medical tomography sytems. For geo-physical, model-based, limited view, and constraint-based tomography, however, ART algorithms are very popular. There are related algebraic techniques which reduce the number of iterations and are less sensitive to noise—these more advanced techniques will not be described here.

4.2.3   Matched Filter Interpretation of Tomography

The reconstruction algorithms described for tomography were introduced using the projection-slice theorem, the idea of filtered backprojection, and iterative algebraic schemes. In contrast to the matched filter approach used to introduce SAR (the correlation receiver as a maximum signal to noise ratio detector), the tomographic techniques represent schemes for inverting a linear operation. Fortunately, there is a simple matched filter interpretation which can be applied to tomographic reconstruction.

As described in Equation 4.2, data in a tomographic system is collected as a set of parallel projections

$$p_\theta(t) = \int_{-\infty}^{\infty} \int_{-\infty}^{\infty} \sigma(x,y)\, \delta\left(x\cos\theta + y\sin\theta - t\right) dxdy. \qquad (4.15)$$

In order to implement a matched filter, a reference signal (the response due to a point image) is required. Suppose that a point image given by a Dirac delta function in polar coordinates as

$$\sigma(r,\phi) = \delta(r - r_0, \phi - \phi_0) \qquad (4.16)$$

is inspected. This will produce the projections

$$p_\theta(t) = \delta(t - r_0\cos(\theta - \phi_0), \theta). \qquad (4.17)$$

As demonstrated in this equation, point images produce sinusoidal traces through the projection data set which leads to the name "sinogram" to describe the collected parallel projection data. Figure 4.10 demonstrates the sinusoidal traces for a simple part with three flaws. The spatial location of the point in the part determines the amplitude (by radial position of

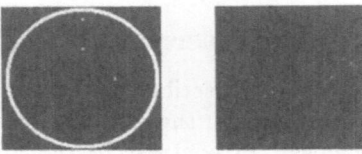

Figure 4.10: Sinogram of projection data from a simple part.

the point) and phase (by angular positon of the point) of the trace in
the sinogram. The appropriate matched filter operation is to integrate
(average for discrete implementations) along the correct sinusoidal trace
in the Radon (projection) domain. The matched filtering operation for a
reference point at polar coordinates (r, ϕ) is therefore given by

$$\hat{\sigma}_{mf}(r, \phi) = \int_0^\pi p_\theta(t) \, \delta[t - r\cos(\theta - \phi), \theta] \, d\theta \qquad (4.18)$$

By eliminating the delta function, this integral can be reduced to

$$\hat{\sigma}_{mf}(r, \phi) = \int_0^\pi p_\theta[r\cos(\theta - \phi)] \, d\theta \qquad (4.19)$$
$$= \hat{\sigma}_{bp}(r, \phi)$$

This simple derivation implies that the matched filter produces the same
reconstruction as the backprojection operation!

As was discussed earlier, there will be an inherent distortion associated
with the backprojection summation reconstruction which can be corrected
by rho-filtering. Because of the equivalence of matched filtering and backprojection summation, the matched filter reconstruction will also require

Figure 4.11: Backprojection summation for three flaw part.

rho-filtering. Figures 4.11 through 4.13 show the backprojection summation and filtered backprojection reconstructions of a part with three flaws. Note the $|\pi r|^{-1}$ tent centered about the flaws in the reconstructions with no rho-filtering.

4.2.4 Summary of Tomographic Techniques

Tomography is a method of imaging a slice through a three-dimensional object. The projection-slice theorem relates the Fourier transform of the original image to the Fourier transform of the parallel projection measurement. This is a powerful analytic tool but requires interpolation in the frequency domain for implementation. Summation backprojection is an alternative algorithm which is relatively efficient to implement on a digital computer but has an inherent distortion of the original image. Because the summation backprojection process is linear, the inherent distortion can be corrected by linear filtering the projection data before backprojection. Matched filter techniques (reconstruction based on correlating the measured data with the expected response due to point images) are equivalent to backprojection summation and require the same rho-filters for correcting the inherent distortion. When a restricted number of projections are available, alge-

Figure 4.12: Filtered backprojection for three flaw part.

No   Filtering

Rho-filtered   with
Rectangular   Window

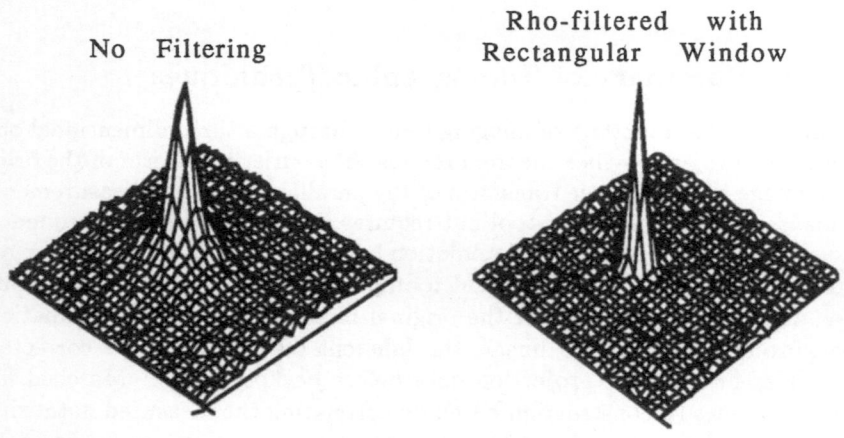

Figure 4.13: Perspective plot of one flaw in the reconstruction.

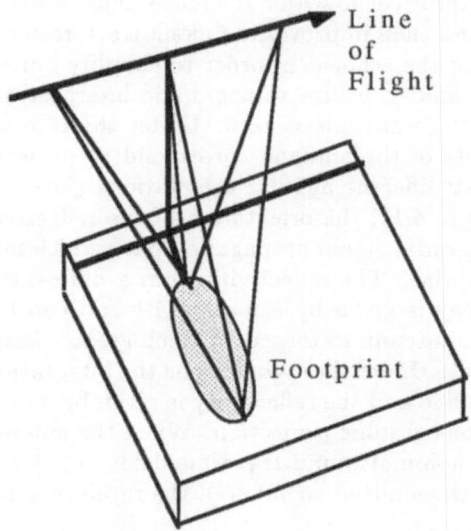

Figure 4.14: Spotlight mode SAR geometry.

braic techniques can incorporate models for the physical imaging process to provide more accurate reconstructions. The relative merits of Fourier inversion, filtered backprojection, and algebraic reconstruction techniques depend on the application and implementation requirements. Because of the wide use of tomographic systems these areas have received extensive research efforts.

4.3 Spotlight Mode SAR

High resolution imagery is reconstructed in strip map SAR by illuminating the target swath with beams that have overlapping footprints. Individual targets are therefore illuminated many times allowing information about their relative azimuthal positions to be extracted from the collection of range echoes. The same type of imaging can be performed by steering a side-looking antenna onto one target area as the platform moves by the region of interest. As shown in Figure 4.14, the antenna is initially forward-looking, incrementally rotates to broadside, and finishes in a backward-looking position. Individual targets in the footprint of interest contribute to essentially all the radar echoes.

As with strip-map SAR, the spotlight radar often transmits a linear fm

signal, mixes the return echo with a reference signal generated using a stable
local oscillator, and then improves the along-track resolution of the system
by post-processing the echoes. In order to simplify our discussion of spotlight mode SAR and to utilize tomographic interpretations, assume that
the antenna platform altitude is zero. Under the zero altitude condition,
the various squints of the antenna correspond to projections of the beam
footprint taken at different angular orientations (just as in tomography).
As shown in Figure 4.15, the orientation of the reflectivity line integral is
orthogonal to the radar signal propagation path which intersects the origin
of the beam footprint. The reflectivity from a differential strip of targets
at equal time delay is given by Equation 4.1 and can be simplified when
the distance from antenna to targets is much greater than the width $2w$ of
the beam footprint. Under these conditions the integration path is approximately a straight line and the reflectivity is given by a line integral which is
equivalent to a tomographic projection. When the antenna interogates the
footprint from a nominal round trip time-delay τ_0, at a perspective angle
of θ, and with a transmitted signal $s(t)$, the received echo is

$$r(t) = \int_{-w}^{w} p_\theta(\tau)\, s(t - \tau - \tau_0)\, d\tau, \tag{4.20}$$

where τ is in round trip time units which can be converted to ground
distances by a $c/2$ scaling (c is the speed of propagation).

As in the strip-mapping SAR case, let the transmitted signal be a linear
fm chirp. Recall, as in Equation 1.16, that the strip-map echo is usually
beaten against a simple sinusoid having a phase equal to the carrier frequency shifted by some convenient intermediate frequency f_{if}. For mixing
the received spotlight echo waveform, however, a linear fm reference pulse
that is matched to the expected echo from a point target at time delay τ_0
is used. The transmitted signal $s(t)$, the echo $r(t)$, and the mixed, low-pass
filtered, and I-Q converted echo $r_{if}(t)$ are given by

$$s(t) = \cos\{j2\pi[f_c t + .5at^2]\}\, \Re_T(t - T/2) \tag{4.21}$$

$$r(t) = \int_{-w}^{w} p_\theta(\tau)\, \cos\{j2\pi[f_c(t - \tau - \tau_0) + .5a(t - \tau - \tau_0)^2]\}\, d\tau$$

$$r_{if}(t) = \int_{-w}^{w} p_\theta(\tau)\, e^{j2\pi .5a\tau^2}\, e^{-j2\pi[f_c + a(t - \tau_0)]\tau}\, d\tau. \tag{4.22}$$

When the quadratic phase term is negligible or when it can be compensated
out of the data, the expression for r_{if} reduces to the Fourier transform
of the projection $p_\theta(t)$. Consequently the projection-slice theorem from
tomography can be invoked to describe spotlight-mode SAR. Typically the

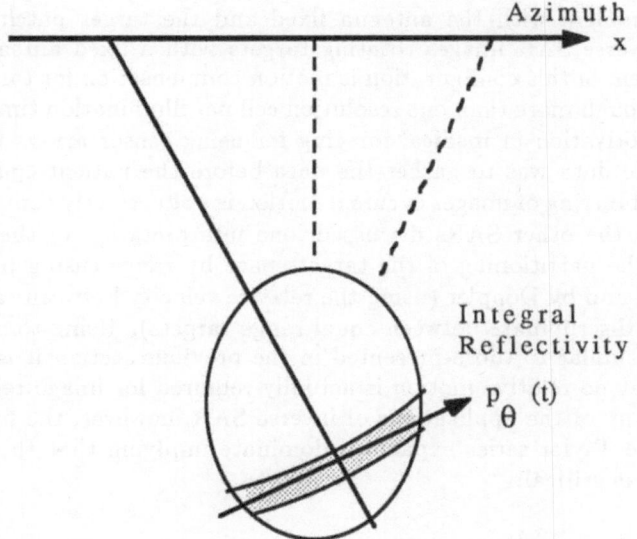

Figure 4.15: Zero altitude spotlight-mode SAR geometry.

flight path of the antenna platform produces a limited number of different viewing angles (less than 180 degrees). By the projection-slice theorem, it is clear that this type of spotlight-mode SAR data will only partially fill the frequency domain. As with limited-view tomography, spotlight SAR algorithms are sensitive to ambiguities due to reconstruction from partial information.

4.4   Inverse SAR

In spotlight-mode SAR a fixed ground patch is imaged by sequentially illuminating the patch from different viewing angles. Inverse SAR is the identical problem with the antenna fixed and the target patch moving. That is, inverse SAR images rotating targets with a fixed antenna. The main problem in this configuration is motion compensation for targets that migrate through more than one resolution cell per illumination time. Recall that the motivation in medical imaging for using sensor arrays to collect tomographic data was to gather the data before the patient could move. Significant blurring of images occurs if motion is not correctly compensated.

As with the other SARs discussed, one interpretation of the imaging process is the partitioning of the target space by range (using pulse echo techniques) and by Doppler (using the relative velocity between target and antenna to discriminate between equal range targets). Using tomographic arguments similar to those presented in the previous section it is possible to show that no relative motion is actually required for image reconstruction. In many of the applications of inverse SAR, however, the first terms of the range Taylor series expansion dominate implying that the Doppler component is critical.

4.5   Summary of Related Algorithms

A brief description of SAFT, tomography, spotlight-mode SAR, and inverse SAR is used to demonstrate some similarities with strip map SAR in the mathematics and the processing. The exchange of ideas between the various disciplines will certainly extend the understanding of these related applications. In addition, the arrival of high speed floating point digital signal processing chips, vector oriented compilers and processors, and multiprocessor computing systems will make the computationally intensive task of reconstructing images possible for an even larger class of applications. Future architectures, such as compilers with access to fast polynomial-based address generators, will make data collection along arcs a quicker operation.

This is important in SAR and SAFT systems where calculating the correct index to gather a vector for processing is the bottleneck in the throughput. Current supercomputers already support the scattered gather for collecting a vector of scattered data at known addresses very quickly and other advances in hardware can be expected.

4.6 References

D. A. Ausherman, A. Kozma, J. L. Walker, H. M. Jones, and E. C. Poggio, "Developments in radar imaging," *IEEE Trans. Aero. and Electronic Syst.*, vol. AES-20, no. 4, pp. 363–400, July 1984.

H. H. Barrett and W. Swindell, "Analog reconstruction methods for transaxial tomography," *Proc. of IEEE*, vol. 65, no. 1, pp. 89–107, Jan. 1977.

L. J. Busse, H. D. Collins, and S. R. Doctor, "Review and discussion of the development of synthetic aperture focusing technique for ultrasonic testing (SAFT-UT)," NUREG/CR-3625, PNL-4957, March 1984.

A. J. Devaney, "A filtered backpropagation algorithm for diffraction tomography," *Ultrasonic Imaging*, vol. 4, pp. 336–350, 1982.

A. C. Kak, "Tomographic imaging with diffracting and nondiffracting sources," in *Array Signal Processing*, ed. S. Haykin, Prentice-Hall, New Jersey, 1985.

D. L. Mensa, S. Halevy, and G. Wade, "Coherent doppler tomography for microwave imaging," *Proc. of IEEE*, vol. 71, no. 2, pp. 254–261, Feb. 1983.

D. C. Munson, *et. al.*, "A tomographic formulation of spotlight-mode synthetic aperture radar," *Proc. of IEEE*, vol. 71, no. 8, pp. 917–925, Aug. 1983.

J. L. Walker, "Range-Doppler imaging of rotating objects," *IEEE Trans. on Aerospace and Electronic Systems*, vol. AES-16, no. 1, pp. 23–51, Jan. 1980.

4.7 Problems

P4.1 How could more sophisticated azimuthal matched filters be used in SAFT?

P4.2 Why is the DC value of the Fourier transform equal for all projections?

P4.3 Manipulate the projection-slice relationship into an expression for filtered backprojection tomography.

P4.4 Why is the integral in Equation 4.8 not over the range zero to 2π?

P4.5 Show that the Fourier transform of $|\pi r|^{-1}$ is $|\rho|^{-1}$.

P4.6 The inverse Fourier transform of a filter with frequency response $H(f) = f$ does not exist. Weight the frequency representation H by a low-pass filter H_L which is a rectangular window and calculate the corresponding impulse response.

P4.7 Using ART, reconstruct a 2 by 2 image from the following six line projections assuming binary weights 0 or 1.

$p_1 = 5$	$p_2 = 10$	$p_3 = 10$	$p_4 = 15$
	σ_1	σ_2	$p_5 = 13$
	σ_3	σ_4	$p_6 = 7$

P4.8 Verify Equation 4.22.

Appendix A

Signal Processing Tools

The fast Fourier transform (FFT) is probably the most powerful tool in digital signal processing. Because of its value as a tool, extensive hardware and software investigations of the FFT have been pursued. These investigations have resulted in many efficient implementations of the FFT. This chapter discusses some fundamental concepts in digital signal processing which culminate in a justification for using the FFT. It should be noted that most theorems, properties, and techniques found in digital signal processing have analogies in the domain of continuous time signals. Therefore many or all of the following discussions will be familiar to readers with a background in linear system theory.

A.1   Sampling Continuous Time Signals

The design of a system to do signal processing begins with a mathematical description of the input signals. The input may be a continuous function of time, such as an AM radio signal, or the input may be a discrete sequence of numbers. Frequently the signal is converted between continuous and discrete time representations through the use of the sampling theorem. This important theorem states that a continuous time signal can be reconstructed from the values it assumes at a certain set of time positions. The time between these samples specifies a rate or frequency at which the continuous signal must be sampled. The sampling theorem states that in order to be able to reconstruct the continuous time input signal from its samples, the rate of taking samples, known as the sampling frequency, must be greater than twice the highest frequency contained in the continuous time input signal. Twice the highest frequency contained in the analog

131

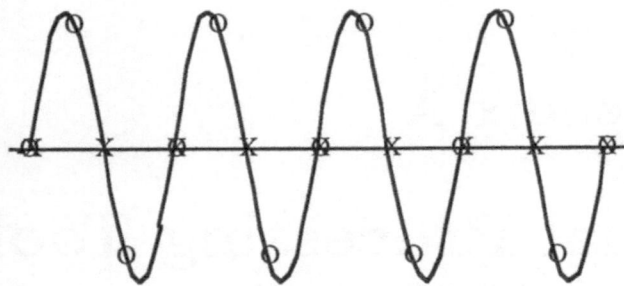

Figure A.1: Sampling a sinusoidal signal at exactly the Nyquist rate (x's) and faster than the Nyquist rate (o's).

signal is often called the Nyquist rate.

The signal $\sin(2\pi f_0 t)$ shown in Figure A.1 provides a simple example of this theorem. The samples denoted by x's were taken at exactly twice the highest frequency in the input—that is, $f_s = 2f_0$ and the conditions of the sampling theorem are not met. Because all of the samples are zero it is clearly impossible to reconstruct the signal from these samples; the continuous time signal could be $\sin(2\pi f_0 t)$ or it could be identically zero. When the sampling frequency f_s is increased such that $f_s > 2f_0$ the resulting samples denoted by o's are now sufficiently close to reconstruct the input $\sin(2\pi f_0 t)$. This is the only sinusoid with frequency less than half the sampling frequency which intersects all of these samples. Note that there is an infinite number of signals with components of frequency greater than half of f_s which intersect all of the samples. For this example, some of these signals can be written

$$\sin[2\pi(f_0 + mf_s)t], \; m = \pm 1, \pm 2, \ldots \qquad (A.1)$$

This can be verified at all the sample positions by replacing the time variable t with an integer multiple of the sampling period. In light of the sampling theorem, when trying to reconstruct a continuous time signal from a sequence of samples, only those signals with components of frequency less than $f_s/2$ are considered.

If during the conversion from continuous time to discrete sequence the sampling theorem is violated, an interesting phenomenon known as aliasing occurs. In simplified terms, the continuous time components which are of frequency greater than half the sampling frequency f_s are translated down to a lower frequency. The frequency information of the resulting sequence can be expressed as the sum of shifted original spectras. Letting X be the frequency representation of the sampled signal and X_a be the frequency information of the continuous time or analog signal x before sampling, then

$$X(f) = \sum_{r=-\infty}^{\infty} X_a(f + rf_s) \qquad (A.2)$$

where X and X_a are defined[1] by

$$X(f) = \sum_{n=-\infty}^{\infty} \bar{x}(n)\, e^{-j2\pi fn} \quad \text{and} \quad X_a(f) = \int_{-\infty}^{\infty} x_a(t)\, e^{-j2\pi ft}\, dt. \quad (A.3)$$

Note that the sequence of points $\bar{x}(n)$ equals $x_a(n/f_s)$ where $1/f_s$ is the time between samples. Equation A.2 gives an appropriate interpretation of aliasing. However, additional scaling terms are useful when describing, graphing, or analyzing aliasing in more detail. In the signal processing literature the convention is to map the analog signal frequency region of $-f_s/2$ to $f_s/2$ into the sequence's spectral region of $-\pi$ to π. An appropriate representation is therefore

$$X(2\pi f/f_s) = f_s \sum_{r=-\infty}^{\infty} X_a(2\pi(f + rf_s)) \qquad (A.4)$$

As seen in Figure A.2, the frequency representation of the sequence is periodic with period 2π. The variable Ω, with $\Omega = 2\pi f$, is often used in place of f for this particular scaling. Note, however, that the $-\pi$ to π convention is not the only one in use. For instance, the unit interval -0.5 to 0.5 is sometimes preferred. Regardless, when f_s is smaller than the Nyquist frequency the spectra overlap, and when the sampling theorem is obeyed no overlap occurs. Aliasing usually represents a non-recoverable error which prevents reconstruction of the original signal from its samples.

It should be noted that reconstructing a continuous time signal from its samples is not always the purpose of a digital signal processing system. However, if it is impossible to reconstruct the analog signal, then there must have been some loss of information during sampling. The information which is not preserved is in the spectral region above half the Nyquist frequency is wrapped around or aliased during sampling which destroys information within the Nyquist bound.

Summarizing, a continuous time signal can be represented as a sequence of sample values without any loss of information as long as the restrictions

---

[1] In order to establish the similarities between the z-transform and the Fourier transform, many texts write $X(f)$ as $X(e^{j\omega})$ which is defined as $\sum \bar{x}(n)e^{-j\omega n}$. Readers unfamiliar with discrete Fourier techniques should be warned that this transform is not the sum calculated by the FFT (see Equation A.24) whose output is a sequence of values.

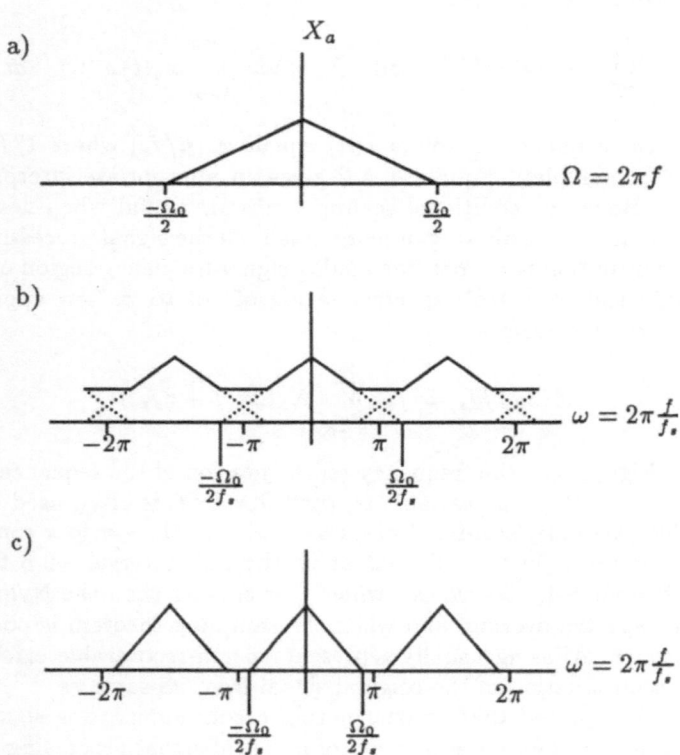

Figure A.2: Frequency effects of sampling an analog signal with spectrum X_a: a) input spectrum; b) aliasing; c) no aliasing.

of the sampling theorem are observed. It should be noted that time was used as the independent variable for the input functions in the previous discussion. When inputs are functions of distance the same sampling rules apply with the Nyquist rate specifying a maximum distance between samples and a maximum spatial frequency in the input.

Radar signals are transmitted and received as continuous functions of time. However, the data is often sampled before being processed. This creates a discrete time sequence and is the reason for using digital signal processing techniques. It is assumed, unless otherwise noted, that the conditions of the sampling theorem were observed during conversion from continuous to discrete time. This results in a mathematical description of the input signals as discrete sequences. These sequences are compactly represented as vectors which are denoted in the text by arrows. For example, the vector x is written $\vec{x}$. The first element of the vector x would be written $\vec{x}(1)$. In signal processing it is often convenient to let the number elements in a vector be infinite. This is done to simplify the mathematics and to allow for periodic signals. The vector (also called an array) is therefore actually a set of elements $\vec{x}(i)$ where i is any integer. When the the vector is known not to be infinite in length the range of values of i is specified. Obviously infinite length vectors can be created from finite length vectors by appending zeros or by considering the finite length vector as the first period of an infinitely long periodic sequence.

A.2 Specifying A Model: Linear Systems

In a radar system a signal is transmitted from an antenna, the electromagnetic energy propagates through space until it is reflected by an object, this reflection propagates back to the antenna which now receives the information. The received signal is then processed to extract particular information about the object(s) which reflected the light waves. This entire process can be treated as a series of mathematical mappings from an input, which represents the transmitted wave, into the desired processed information. This is diagrammed in Figure A.3. The mapping G transforms the system input into the received signal. This corresponds to the physical phenomenon of the transmission, reflection, and reception of the signal. The mapping F takes the received signal (output of operation G) and transforms it into the output of the radar system. The goal is to design an appropriate mapping F which produces a quality estimate of the desired information.

Because there are many well developed and powerful design tools associated with the class of linear mappings, restricting the mapping F to

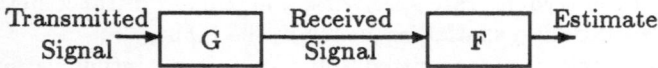

Figure A.3: Representation of a radar system as mathematical mappings.

be within this class reduces the complexity of solving for the best esti-
mator. Representation of a non-linear mapping as a linear mapping will
introduce errors. However, if satisfactory results can be obtained using a
linear representation of the system, a significant savings in design time and
implementation costs can be realized by exploiting the existing tools and
properties of linear mappings.

A mapping L is said to be a linear mapping if for any inputs $\vec{x}$ and $\vec{y}$
and any real numbers a and b

$$L(a\vec{x} + b\vec{y}) = aL(\vec{x}) + bL(\vec{y}) \tag{A.5}$$

This is known as the superposition definition for linear mappings. An
appropriate interpretation is to think of $a\vec{x} + b\vec{y}$ as one single vector $\vec{z}$.
Then

$$L(\vec{z}) = L(a\vec{x} + b\vec{y}) = aL(\vec{x}) + bL(\vec{y})$$

This means that input vectors to linear mappings may be decomposed into
sets of weighted vectors which can be taken through the mapping separately
then recombined. Mathematically, if $\vec{z} = \sum_{i=0}^{N-1} a_i \vec{x}_i$ then

$$L(\vec{z}) = L\left(\sum_{i=1}^{N-1} a_i\vec{x}_i\right) = \sum_{i=0}^{N-1} L(a_i\vec{x}_i) = \sum_{i=0}^{N-1} a_i\, L(\vec{x}_i) \tag{A.6}$$

Note that $\vec{x}_i$ denotes the $i'th$ vector from a set of vectors—not the $i'th$
element of $\vec{x}$ which is written $\vec{x}(i)$. The tools and properties of linear
systems are all based on this definition.

A constraint known as shift invariance is also frequently imposed in
addition to the linear restriction. Suppose that $\vec{z}_n(m) = \vec{x}(m - n)$ for all
m and any n—that is, $\vec{z}_n$ is $\vec{x}$ shifted n positions. Denote the output of
the mapping T applied to each of these two vectors by $\vec{Z}_n = T(\vec{z}_n)$ and
$\vec{Z} = T(\vec{x})$. The mapping T is said to be shift invariant if for every n

$$\vec{Z}_n(m) = \vec{Z}(m - n), \qquad \text{for all } m. \tag{A.7}$$

That is, a shift in the input sequence results in an identical shift in the output. Systems with both the above properties are called linear shift-invariant (LSI) systems. When the indices represent the parameter time, these are commonly referred to as linear time-invariant systems.

In this paper, both the signal processing tools and the models constructed for the mappings will assume a linear shift-invariant system. It is important to recognize that the physical phenomenon associated with a particular application may not have either of these properties. This must be considered when evaluating the performance of the processing system and when proposing modifications to the model. For now, the linear time invariant model will be used to develop and explain several signal processing techniques.

A.3 The Convolution Representation

The superposition definition provides the mathematics necessary to develop a useful representation for linear systems. Taking an input $\vec{x}$ and decomposing it into a set of weighted vectors can be done in numerous ways. The following approach is based on a vector known as the unit sample function. This vector is denoted as $\vec{\delta}$ and is defined by

$$\vec{\delta}(m) = \left\{ \begin{array}{ll} 1, & \text{if } m = 0; \\ 0, & \text{else.} \end{array} \right. \tag{A.8}$$

Readers who are familiar with continuous time linear systems will recognize the similarity between the unit sample vector and the Dirac delta function.

The output of a linear shift-invariant mapping L when $\vec{\delta}$ is the input vector is called the unit sample response. Let $\vec{h}$ denote this response, that is $\vec{h} = L(\vec{\delta})$. Because the system is shift invariant, the vector $\vec{h}$ characterizes the output of L for any shifted unit sample input. In order to explain this, let

$$\vec{\delta}_n(m) = \left\{ \begin{array}{ll} 1, & \text{if } m = n; \\ 0, & \text{else.} \end{array} \right.$$

Because $\vec{\delta}(m - n) = \vec{\delta}_n(m)$ for all m, it is simply a shifted version of our unit sample vector. Using $\vec{h}_n = L(\vec{\delta}_n)$ to represent the response of L to $\vec{\delta}_n$, it is clear from the shift invariance property of the system that for every n

$$\vec{h}_n(m) = \vec{h}(m - n), \qquad \text{for all } m.$$

This is what is meant by $\vec{h}$ completely characterizing the system for all unit sample inputs.

The next step is to use superposition and the unit sample responses to determine the output of the system for an arbitrary input $\vec{x}$. The input vector $\vec{x}$ can be expressed as a weighted sum of the shifted unit sample vectors as follows

$$\vec{x} = \sum_n \vec{x}(n)\, \vec{\delta}_n \qquad (A.9)$$

Note that $\vec{x}(n)$ is the $n'th$ element of $\vec{x}$ and can be treated as a real number in this equation. It assumes the same role as a_i in Equation A.6. Using linearity to calculate the mapping for this representation yields

$$
\begin{aligned}
\vec{Z} &= L(\vec{x}) \\
&= L\left(\sum_n \vec{x}(n)\, \vec{\delta}_n \right) \\
&= \sum_n \vec{x}(n)\, L(\vec{\delta}_n) \\
&= \sum_n \vec{x}(n)\, \vec{h}_n \qquad (A.10)
\end{aligned}
$$

Exploiting the shift invariant system assumption, implies that the $m'th$ element of this response to $\vec{x}$ is

$$
\begin{aligned}
\vec{Z}(m) &= \sum_n \vec{x}(n)\, \vec{h}_n(m) \\
&= \sum_n \vec{x}(n)\, \vec{h}(m-n) \qquad (A.11)
\end{aligned}
$$

Equation A.11 is known as the convolution sum. Summarizing this approach: given a linear shift invariant mapping L with an associated unit sample response $\vec{h}$ and any input $\vec{x}$, the output $\vec{Z} = L(\vec{x})$ is specified by the convolution of $\vec{x}$ with $\vec{h}$. Because the unit sample response is sufficient for calculating the output of the linear shift invariant mapping L for any input $\vec{x}$, this convolution is referred to as linear convolution and is denoted $\vec{x} * \vec{h}$. Any linear shift invariant system can be represented this way. Because calculating the convolution sum for every desired output position may be computationally tedious, a fast algorithm for performing convolution is desired.

A.4   Fast Convolution with FFTs

The Fourier series representation of a continuous time function $f(t)$ on an interval of length l is given by

$$f(t) = \sum_{k=-\infty}^{\infty} c_k\, e^{jk\pi t/l}. \tag{A.12}$$

The letter j is used to denote the imaginary number $\sqrt{-1}$. By Euler's Law

$$e^{j\phi} = \cos(\phi) + j\sin(\phi). \tag{A.13}$$

Consequently the argument ϕ can be reduced modulo 2π without changing the value of $e^{j\phi}$. This periodicity of $e^{j\phi}$ means that the discrete Fourier series (DFS) representation of a sequence can be reduced to a finite number of terms whenever the sequence being represented is periodic. The convention for this paper is to denote vectors that are known to be periodic by a tilde "$\sim$". The vector $\tilde{x}$ is said to be periodic with period N if for any n

$$\tilde{x}(n + rN) = \tilde{x}(n), \qquad r = \pm 1, \pm 2, \ldots \tag{A.14}$$

The Fourier series representation of this sequence is given by

$$\tilde{x}(n) = \frac{1}{N} \sum_{k=0}^{N-1} c_k\, e^{j\,2\pi nk/N}. \tag{A.15}$$

Because $\tilde{x}$ and $e^{j\,2\pi nk/N}$ are periodic with period N the sum need only be over a finite range. To solve for the coefficients c_k, calculate

$$
\begin{aligned}
\sum_{n=0}^{N-1} \tilde{x}(n)\, e^{-j2\pi nm/N} &= \frac{1}{N} \sum_{n=0}^{N-1}\sum_{k=0}^{N-1} c_k\, e^{-j2\pi n(m-k)/N} \\
&= \sum_{k=0}^{N-1} c_k \left[\frac{1}{N} \sum_{n=0}^{N-1} e^{-j2\pi n(m-k)/N} \right]
\end{aligned}
$$

The term in the brackets is zero except when $m - k$ is an integer multiple of N. For $m = k \pm rN$ and any integer r, the exponential is one, the sum is N and the entire expression reduces to $c_{k\pm rN}$. Therefore

$$c_{k\pm rN} = \sum_{n=0}^{N-1} \tilde{x}(n)\, e^{-j2\pi nk/N}$$

for any integer r. That is, the coefficients form a periodic sequence. We write this periodic sequence of Fourier coefficients as $\tilde{X}$ where $\tilde{X}(k) = c_k$. The pair of equations for the discrete Fourier series representation of a periodic sequence is therefore

$$\tilde{x}(n) = \frac{1}{N} \sum_{k=0}^{N-1} \tilde{X}(k) \, e^{j2\pi nk/N} \qquad (A.16)$$

$$\tilde{X}(k) = \sum_{n=0}^{N-1} \tilde{x}(n) \, e^{-j2\pi nk/N} \qquad (A.17)$$

Our reason for examining Fourier representations of sequences is to find a fast algorithm for calculating convolutions. Not by coincidence, there is a convolution type operation which results from taking the product of two DFS representations of periodic sequences. In order to analyze this, let $\tilde{X}_1$ and $\tilde{X}_2$ be the DFS representations of the periodic sequences $\tilde{x}_1$ and $\tilde{x}_2$, respectively. Because $\tilde{x}_1$ and $\tilde{x}_2$ are both of period N, so will $\tilde{X}_1$, $\tilde{X}_2$, and their product which we denote as $\tilde{Z}$. Then

$$
\begin{aligned}
\tilde{z}(n) &= \frac{1}{N} \sum_{k=0}^{N-1} \tilde{Z}(k) \, e^{j2\pi nk/N} \\
&= \frac{1}{N} \sum_{k=0}^{N-1} \tilde{X}_1(k) \, \tilde{X}_2(k) \, e^{j2\pi nk/N} \\
&= \frac{1}{N} \sum_{k=0}^{N-1} \sum_{m=0}^{N-1} \sum_{r=0}^{N-1} \tilde{x}_1(m) \, \tilde{x}_2(r) \, e^{-j2\pi(m+r-n)k/N} \\
&= \sum_{m=0}^{N-1} \sum_{r=0}^{N-1} \tilde{x}_1(m) \, \tilde{x}_2(r) \left[\frac{1}{N} \sum_{k=0}^{N-1} e^{-j2\pi(m+r-n)k/N} \right] \qquad (A.18)
\end{aligned}
$$

The quantity in the brackets is zero unless $m + r - n$ is an integer multiple of N. When $m + r - n$ is an integer multiple of N, or zero, the bracket reduces to unity. This implies that

$$\tilde{z}(n) = \sum_{m=0}^{N-1} \tilde{x}_1(m) \, \tilde{x}_2(n-m). \qquad (A.19)$$

Note the similarity between this result and the linear convolution of Equation A.11. Because the sum is over one period instead of all positions in the vectors, this is called periodic convolution. This is not surprising since one period of a periodic sequence is sufficient for describing the entire sequence.

How to use periodic convolution and periodic sequences to represent linear systems with nonperiodic inputs will be discussed next.

Inputs and unit sample responses will typically be data sequences or computer generated arrays. Consequently most practical signals can be considered of some finite length N. For convenience, and by convention, positions 0 through $N-1$ of the vector are considered as the signal. The finite length signal can be thought of as an infinitely long vector padded with zeros outside of this region. In order to have all vectors the same length, these infinitely long vectors created by appending zeros will be considered as "finite length signals." However, in order to use the results of the discrete Fourier series representation of periodic signals, it is convenient to consider the length N signal as the first period from an infinitely long periodic sequence. For positions 0 through $N-1$ these two interpretations are identical. Assuming the signal to be one period of a periodic signal, however, will produce several interesting interpretations of the Fourier operations.

To simplify conversions between finite length signals and their periodic representation, it is convenient to define $\Re_N$ as the rectangle function where

$$\Re_N(n) = \left\{ \begin{array}{ll} 1, & \text{if } 0 \leq n \leq N-1; \\ 0, & \text{else,} \end{array} \right. \tag{A.20}$$

and the modulo operator $\mod(n, N) = n - rN$ where the integer r is selected such that $n - rN$ is between 0 and $N-1$. The vector $\vec{x}$ which is zero outside of positions 0 through $N-1$ can be represented as the periodic sequence $\tilde{x}$ where

$$\tilde{x}(n) = \sum_{r=-\infty}^{\infty} \vec{x}(n + rN) = \vec{x}(\mod(n, N)); \tag{A.21}$$

$$\vec{x}(n) = \tilde{x}(n)\,\Re_N(n). \tag{A.22}$$

Interpreting finite length signals as periodic signals results in several unusual properties. An important example is the results of shifting a vector. For the periodic interpretation, shifting a vector implies shifting the periodic representation of the vector and then taking the values in positions 0 through $N-1$ as the finite length shifted vector. Let $\vec{z}$ be the vector $\vec{x}$ shifted m positions. Then

$$\begin{aligned} \vec{z}(n) &= \tilde{x}(n-m)\,\Re_N(n) \\ &= \vec{x}(mod(n-m, N))\,\Re_N(n). \end{aligned} \tag{A.23}$$

As the signal $\vec{x}$ shifts, in this case to the right when m is positive, the terms which are shifted outside of the window "wrap around" and appear at the

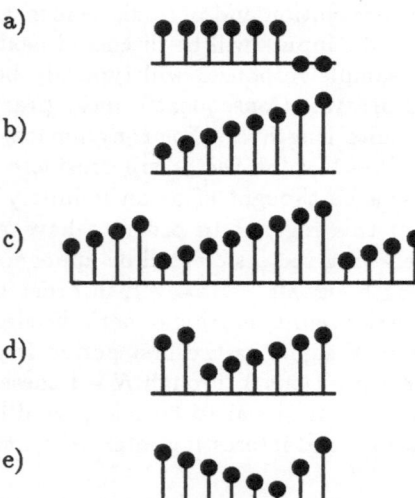

Figure A.4: Effects of representing finite length signals as one cycle of a periodic sequence: a) and b) are input signals; c) is the periodic representation of b; d) is the circular shift of b) two positions to the right; e) is the circular convolution of signals a) and b).

other end of the window. The first N values in $\vec{x}$ can be thought of as forming a circle. This circle then defines the first period of an infinitely long sequence. A shift of the sequence $\vec{x}$ is therefore a rotation of the circle. This interpretation gives the name "circular shift" to these operations. An example of circular shifting is given in Figure A.4.

In keeping with the interpretation of finite length signals as one period of an infinite signal, define the Discrete Fourier Transform (DFT) pair as $\vec{x}$ and $\vec{X}$ where

$$\vec{X}(k) = \sum_{n=0}^{N-1} \vec{x}(n)\, e^{-j2\pi kn/N} \qquad (A.24)$$

$$\vec{x}(n) = \frac{1}{N} \sum_{k=0}^{N-1} \vec{X}(k)\, e^{+j2\pi kn/N} \qquad (A.25)$$

The Fast Fourier Transform (FFT) denotes a class of algorithms which are used to calculate the Discrete Fourier Transform efficiently. Improvements in computation speed and memory use are possible with the FFT. This improved efficiency makes the FFT an important tool.

Unlike their periodic counterparts in Equations A.16 and A.17, the DFT pair are always zero outside of the first N positions. The relationship between these two Fourier representations is given by

$$\vec{X}(k) = \tilde{X}(k)\,\Re_N(k) \ \text{ and } \ \vec{x}(n) = \tilde{x}(n)\,\Re_N(n). \qquad (\text{A.26})$$

Just as with the periodic sequences, a convolution type operation can be performed by the products of the DFT's. To examine the convolution type operation, let $\vec{X}_1$ and $\vec{X}_2$ be the DFT's of the finite length signals $\vec{x}_1$ and $\vec{x}_2$, respectively. Also, let $\vec{Z}$ be the product of $\vec{X}_1$ with $\vec{X}_2$. Then

$$
\begin{aligned}
\vec{z}(n) &= \frac{1}{N} \sum_{k=0}^{N-1} \vec{Z}(k)\, e^{j2\pi nk/N} \\[2mm]
&= \frac{1}{N} \sum_{k=0}^{N-1} \vec{X}_1(k)\, \vec{X}_2(k)\, e^{j2\pi nk/N} \\[2mm]
&= \frac{1}{N} \sum_{k=0}^{N-1} \tilde{X}_1(k)\, \tilde{X}_2(k)\, e^{j2\pi nk/N}\, \Re_N(k) \qquad (\text{A.27})
\end{aligned}
$$

Using the same approach as in the derivation of EquationA.19, this can be reduced to the first N terms of the periodic convolution of $\tilde{x}_1$ and $\tilde{x}_2$.

$$
\begin{aligned}
\vec{z}(n) &= \sum_{m=0}^{N-1} \tilde{x}_1(m)\, \tilde{x}_2(n-m)\, \Re_N(n) \qquad (\text{A.28}) \\[2mm]
&= \sum_{m=0}^{N-1} \vec{x}_2(\text{mod}(m,N))\, \vec{x}_2(\text{mod}(n-m,N))\, \Re_N(n) \quad (\text{A.29})
\end{aligned}
$$

This final equation performs an operation very similar to convolution. Because the vectors are circularly shifted, multiplied, and summed, this is known as circular convolution. Equation A.28 clearly shows that the circular convolution of $\vec{x}_1$ and $\vec{x}_2$, denoted $\vec{x}_1 \otimes \vec{x}_2$, is equivalent to taking the first period of the periodic convolution of $\tilde{x}_1$ and $\tilde{x}_2$. This is consistent with the interpretation that finite length signals represent one period of an infinitely long periodic signal. Note that by a simple change of indices it can be shown that $\vec{x}_1 \otimes \vec{x}_2$ equals $\vec{x}_2 \otimes \vec{x}_1$. This symmetric property of circular convolution is also true for linear convolution. In fact, with slight modifications circular convolution can be used to implement linear convolutions.

Suppose the linear convolution of a length N_1 sequence $\vec{x}_1$ with a length N_2 sequence $\vec{x}_2$ is desired. If $\vec{x}_3 = \vec{x}_1 * \vec{x}_2$, then by Equation A.11

$$\vec{x}_3(m) = \sum_{n=-\infty}^{\infty} \vec{x}_1(n)\,\vec{x}_2(m-n). \qquad (A.30)$$

There will be $N_1 + N_2 - 1$ terms where the finite length regions of the two vectors overlap. Before this convolution can be done using DFT's, a common length is needed for the finite length vectors. Because we know the linear convolution could have as many as $N_1 + N_2 - 1$ nonzero terms, treat $\vec{x}_1$ and $\vec{x}_2$ as if they were of length N_3 where $N_3 = N_1 + N_2 - 1$ and the additional values are set to zero. This procedure is descriptively called zero padding. Performing the circular convolution of these length N_3 vectors results in a linear convolution of $\vec{x}_1$ and $\vec{x}_2$ because only the zeros of the longer sequences "wrap around". This can be seen in Figure A.4.

A.5 Summary of Signal Processing Tools

All linear shift invariant systems can be described by their unit sample response: the output of a linear system is the convolution of its unit sample response with the input. The Discrete Fourier Transform (DFT) pair provides a means of doing circular convolution. The Fast Fourier Transform (FFT) is an algorithm which makes the implementation of DFT's, and consequently circular convolutions, very fast. Finally, linear convolutions can be performed by FFT's using the circular convolution of zero padded inputs. Consequently for any linear shift invariant system, the FFT represents a fast method for calculating the response of the system to an input. Because linear models are effective in radar processing, the FFT is a powerful tool.

A.6 References

A. V. Oppenheim and R. W. Schafer, *Digital Signal Processing*, Prentice-Hall, Inc., New Jersey, 1975.

A.7 Problems

PA.1 Compute the discrete Fourier transform (DFT) X of the length 4 sequence x, where $x(0) = 1$, $x(1) = 1$, $x(2) = 0$, and $x(3) = 0$.

PA.2 Compute the length 4 frequency domain sequence Z defined by $Z(m) = |X(m)|^2$, where X is from Problem 1. Then calculate the inverse discrete Fourier transform z.

PA.3 Compute (after zero padding) the autocorrelation of the sequence x in Problem 1 (could use the linear convolution of the sequence x with a time-reversed version of itself). Compare the output of this operation with the results of Problem 2 above.

PA.4 Repeat Problem 2 with $Z(m) = X(m)^2$. Then show, as in Problem 3, that the linear convolution of x with itself equals z.

Appendix B

Matched Filter Derivation

One model for designing a decision procedure to test the hypothesis that a certain deterministic signal occurs in a received waveform assumes additive noise, a linear processor, and an optimization criteria which maximizes the signal to noise ratio at some time instant. Under these assumptions, an integral equation which specifies the impulse response for the optimal detection filter is derived—the result is known as the matched filter integral equation. A filter for the special case of white noise is designed as an example.

B.1 Problem Definition

The input $X(t)$ to a linear filter h is assumed to be the sum of two functions $s(t)$ and $N(t)$. The signal $s(t)$ is some deterministic and known waveform and the noise $N(t)$ is a function which describes the random variations (in a statistical sense) added to the waveform. The function $N(\cdot)$ is referred to as a stochastic process with $N(t)$ being a random variable for each fixed t. This model is illustrated in Figure B.1. Typically the filtering operation is followed by comparison with a threshold to decide for or against the hypothesis that the waveform s is embedded in the signal X.

The linear system h is to be designed so that the output signal to noise ratio (SNR) at time instant $t = T$ is a maximum. This can be expressed as

$$\mathrm{SNR}(t) = \frac{s_0^2(t)}{E\{N_0^2(t)\}} \text{ and } \mathrm{SNR}_{max} = \mathrm{SNR}(T) = \frac{s_0^2(T)}{E\{N_0^2(T)\}}. \quad \text{(B.1)}$$

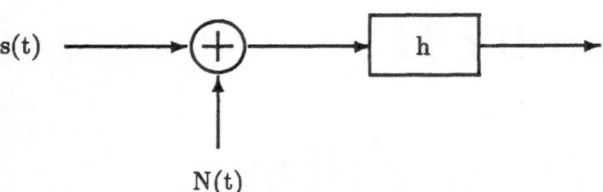

Figure B.1: Model for matched filter derivation.

The notation $E\{\cdot\}$ is used to denote the statistical expectation (average) operator. The convolutional representation of the output signal and the output noise, respectively, are

$$s_0(T) = \int_{-\infty}^{T} s(\tau)\, h(T - \tau)\, d\tau$$

$$N_0(T) = \int_{-\infty}^{T} N(\tau)\, h(T - \tau)\, d\tau \qquad (B.2)$$

Consequently, the output signal to noise ratio at time T can be written

$$\mathrm{SNR}(T) = \frac{|\int_{-\infty}^{T} s(\tau)\, h(T - \tau)\, d\tau|^2}{\int_{-\infty}^{T} \int_{-\infty}^{T} \mathrm{R}(\tau, \sigma)\, h(T - \tau)\, h(T - \sigma)\, d\tau\, d\sigma} \qquad (B.3)$$

where the function R is the autocorrelation of the noise process. The autocorrelation is defined by the statistical expectation operator E with

$$\mathrm{R}(\tau, \sigma) = E\{N(\tau)\, N(\sigma)\}. \qquad (B.4)$$

In the next section, maximization of the SNR at time T is shown to imply a filter shape h which satisfies the matched filter integral equation

$$\int_{-\infty}^{T} \mathrm{R}(\tau, \sigma)\, h(T - \sigma)\, d\sigma = s(\tau), \quad -\infty < \tau \leq T. \qquad (B.5)$$

Given an autocorrelation for the noise process, it is possible to solve this equation for the optimal filter. The derivation of the integral requirement for optimality is included for completeness.

B.2 Optimization: Maximizing the SNR

Our goal is to derive an integral equation describing the optimal (in the maximum SNR sense) linear filter using the calculus of variations. For simplicity let the maximum value attained by the function $\text{SNR}(t)$ be the constant $1/\alpha$. Then by definition

$$\text{SNR}(t) = \frac{s_0^2(t)}{E\{N_0^2(t)\}} \leq \frac{1}{\alpha}$$

or

$$E\{N_0^2(t)\} - \alpha\, s_0^2(t) \geq 0. \tag{B.6}$$

The optimum filter maximizes the SNR which implies equality holds in the previous equation. For our design scheme, this occurs when $t = T$. Expanding out the terms

$$\begin{aligned}
&\int_{-\infty}^{T}\int_{-\infty}^{T} R(\tau,\sigma)\, h(T-\tau)\, h(T-\sigma)\, d\tau d\sigma \\
&\quad - \alpha \left| \int_{-\infty}^{T} s(\tau)\, h(T-\tau)\, d\tau \right|^2 \geq 0.
\end{aligned} \tag{B.7}$$

Because the optimum filter is achieved when equality occurs, the optimization can be performed by minimizing the above expression. The calculus of variations is used for this purpose. If $\delta h(t)$ is used to represent an arbitrary variation about $h(t)$, then the previous constraints on the SNR can be rewritten as

$$\begin{aligned}
&\int_{-\infty}^{T}\int_{-\infty}^{T} R(\tau,\sigma)\, [h(T-\tau) + \epsilon\, \delta h(T-\tau)]\, [h(T-\sigma) + \epsilon\, \delta h(T-\sigma)]\, d\tau d\sigma \\
&\quad - \alpha \left| \int_{-\infty}^{T} s(\tau)\, [h(T-\tau) + \epsilon\, \delta h(T-\tau)]\, d\tau \right|^2 \geq 0.
\end{aligned} \tag{B.8}$$

Collecting terms in the form of a polynomial of the real valued variable ϵ results in

$$P(\epsilon) = P_0 + 2\epsilon P_1 + \epsilon^2 P_2 \tag{B.9}$$

where

$$\begin{aligned}
P_1 &= \frac{1}{2}\int_{-\infty}^{T} \delta h(T-\tau) \int_{-\infty}^{T} R(\tau,\sigma)\, h(T-\sigma)\, d\sigma d\tau \\
&\quad + \frac{1}{2}\int_{-\infty}^{T} \delta h(T-\sigma) \int_{-\infty}^{T} R(\tau,\sigma)\, h(T-\tau)\, d\tau d\sigma \\
&\quad - \frac{\alpha}{2}\int_{-\infty}^{T} s(\sigma)\, h(T-\sigma)\, d\sigma \int_{-\infty}^{T} s(\tau)\, \delta h(T-\tau)\, d\tau \\
&\quad - \frac{\alpha}{2}\int_{-\infty}^{T} s(\tau)\, h(T-\tau)\, d\tau \int_{-\infty}^{T} s(\sigma)\, \delta h(T-\sigma)\, d\sigma
\end{aligned} \tag{B.10}$$

Extrema of the polynomial $P(\epsilon)$ can be found by

- Noting that $\epsilon = 0$ is identical to the equations with no variation about the optimal impulse response;

- Taking the first partial derivative of P with respect to ϵ and equating with zero will describe extrema of the polynomial P.

The result of both these operations is

$$\frac{dP(\epsilon)}{d\epsilon}\Big|_{\epsilon=0} = P_1 = 0.$$

Recall that $s_0(T)$ is used to denote the component of the output signal at time $t = T$ due to the known deterministic input signal. Using the definition of $s_0(T)$, the expression for $P_1 = 0$ can be reduced to

$$\int_{-\infty}^{T} \delta h(T-\tau) \left\{ \int_{-\infty}^{T} R(\tau,\sigma)\, h(T-\sigma)\, d\sigma - \alpha\, s_0(T)\, s(\tau) \right\} d\tau = 0. \quad \text{(B.11)}$$

Because $\delta h(t)$ is arbitrary, the argument in the brackets is required to be identically zero. This implies

$$\int_{-\infty}^{T} R(\tau,\sigma)\, h(T-\sigma)\, d\sigma - \alpha\, s_0(T)\, s(\tau) = 0, \quad -\infty < \tau \le T,$$

or

$$\int_{-\infty}^{T} R(\tau,\sigma)\, h(T-\sigma)\, d\sigma = \alpha\, s_0(T)\, s(\tau), \quad -\infty < \tau \le T.$$

Note that $\alpha\, s_0(T)$ is a constant which causes a magnitude scale change but does not change the filter shape or the maximization of the SNR. For convenience, set $\alpha\, s_0(T) = 1$ which implies

$$\int_{-\infty}^{T} R(\tau,\sigma)\, h(T-\sigma)\, d\sigma = s(\tau), \quad -\infty < \tau \le T. \quad \text{(B.12)}$$

The solution of this equation is the impulse response of the causal filter $h(t)$ that maximizes the SNR at time $t = T$. If the noise is wide sense stationary (wss), then the equation can be solved with a time invariant $h(t)$. The noise random process is wide sense stationarity if its autocorrelation is a function of the difference of the two time arguments alone. That is,

$$R(\tau,\sigma) = R(\tau - \sigma). \quad \text{(B.13)}$$

If a time varying $h(t,\sigma)$ were allowed, then solutions exist for for more general non-stationary noises and the wss assumption is not required. Filters which are optimum in this sense are known as matched filters. The following example demonstrates how this name applies.

B.3 Example Assuming White Noise

The impulse response for the matched filter assuming a white noise source is derived from the integral equation. The autocorrelation for white noise is

$$R(\tau, \sigma) = \frac{N_0}{2} \delta(\tau - \sigma).$$

Note that δ in this context corresponds to the Dirac delta or impulse function. Using this in the matched filter equation implies

$$s(\tau) = \int_{-\infty}^{T} R(\tau, \sigma) h(T - \sigma) \, d\sigma, \quad -\infty < \tau \leq T \quad \text{(B.14)}$$

$$= \int_{-\infty}^{T} \frac{N_0}{2} \delta(\tau - \sigma) h(T - \sigma) \, d\sigma \quad \text{(B.15)}$$

$$= \frac{N_0}{2} h(T - \tau) \quad \text{(B.16)}$$

Using the change of variables $T - \tau = t$:

$$h(t) = \frac{2}{N_0} s(T - t)$$

Because scaling constants can be ignored, we conclude that for white noise, the filter shape which maximizes the signal to noise ratio at time $t = T$ is given by the impulse response

$$h(t) = s(T - t). \quad \text{(B.17)}$$

Note that this is simply a time reversed version of the known deterministic input signal without noise. The filter impulse response is "matched" to the expected input signal shape in the absence of noise.

B.4 Summary of Matched Filter Statistics

The matched filter integral equation specifies the optimal linear filter for maximizing the signal to noise ratio at a single time assuming a known and deterministic signal combined with additive noise. A threshold test can be performed to decide (with some statistical certainty) between the presence or absence of the signal in a particular waveform. Adjustments to the theory allow for many known signals in additive noise. The detection problem then becomes one of selecting which signals are most likely to be in the corrupted data.

B.5   References

W. B. Davenport and W. L. Root, *Random Signals and Noise*, McGraw-Hill, New York, 1958.

E. Wong, *Intro. to Random Processes*, Springer-Verlag, New York, 1983.

B.6   Problems

PB.1 Complete the algebra necessary to reduce the expression $P_1 = 0$ to Equation B.11.

PB.2 Express the transfer function for the matched filter in white noise in terms of $S(\omega)$, the Fourier transform of s.

PB.3 The autocorrelation of a deterministic signal x is often written as $A_x(\tau)$ where

$$A_x(\tau) = \int_{-\infty}^{\infty} x(t)\, x(t+\tau) dt.$$

Explain why the matched filter for white noise is also called the correlation receiver.

PB.4 If complex valued signals (and noise) are being processed, what filter impulse response would be used in white noise?

PB.5 Design a linear detection filter (transfer function), which maximizes the signal to noise ratio at time $t = T$ in the presence of additive wide sense stationary (wss) noise. Assume that a non-causal filter can be used.

Appendix C

Matched Filter Implementation

The material contained in this appendix is not essential in a first reading of the body of the notes. In fact, many of the algorithms merit investigation only when a large amount of repetitive processing is to be performed. The cost of developing and testing these more complicated schemes may not be justified by the performance improvement they provide. As a positive comment: readers may find that these discussions on the relatively simple and standard modifications to the FFT algorithm provide additional insight into the fundamentals of the algorithm as well as signal processing in general.

The relative lengths of the reference waveform and the sampled data would typically determine the best software implementation of the filtering operation. Specifically, short reference sequences are amenable to convolution implementations, where shifts and adds are performed to accomplish the linear filtering. As the reference length approaches the data length, the efficiency of the fast Fourier transform (FFT) filter implementation is increased. However, additional considerations such as the availability of array processors, special purpose hardware, efficient microcoded processing algorithms, as well as the amount of data to be processed, and the cost of implementing new algorithms influence the strategy for each system. In some systems, reliability and expandability may be more important than maximum efficiency. The following section presents some elementary software modifications which can improve the efficiency of the radar processor. For these notes, particular attention is paid to the fast Fourier transform (FFT) algorithm.

The typical radar signal processor will perform the following operations on the sampled received signal $\vec{r}$

1. Appropriately sample the reference signal $u(t)$ to obtain $\vec{u}$.

2. Take the FFT of the sequence $\vec{u}$ creating $\vec{U}$.

3. Complex conjugate $\vec{U}$ to create the matched filter $\vec{U}^*$.

4. Take the FFT of the sequence $\vec{r}$ creating $\vec{R}$.

5. Implement filter by multiplying $\vec{R}$ with $\vec{U}^*$.

6. Convert back to time sequence with inverse FFT.

In many cases the same reference function can be used on a large number of received signals. Storing the matched filter coefficients generated in step 3 eliminates the need to repeat the first three steps for each iteration of the filter. This is a very intuitive means for increasing the operating speed of the radar processor. There are many other techniques for increased performance that are not as obvious. These techniques usually involve additional control code or modified FFT algorithms.

C.1 Eliminating The Bit Reversal Operation

Typically the filtering operation (a linear convolution) is actually implemented as a frequency domain multiply between $\vec{U}^*$ and $\vec{R}$. If the new sequence created by this operation is called $\vec{Y}$, then $\vec{Y}(k) = \vec{R}(k)\vec{U}^*(k)$ for all appropriate k. This can be thought of as an iterative loop in a program or as a vector multiply on an array processor. The vector multiply calculates the product for all the indices simultaneously; the loop performs the same operation sequentially—one index at a time. The order selected to perform the multiplication is not important as long as all indices are covered upon completion. The loop could run from the first through last position, last through first, or through some random order selected by the programmer. Again, there is no preferred order to performing the multiplication. The only condition is that the vector $\vec{Y}$ is an appropriate input to the inverse FFT algorithm used for conversion back to the time domain. This independence of the sequence $\vec{Y}$ from a preferred order of performing the multiplications can be utilized in FFT processing.

Most FFT and inverse FFT (IFFT) algorithms are actually performed in two steps and often by two different computer routines. The first step calculates the discrete Fourier transform values. This calculation is usually

Table C.1: Calculation of bit reversed indices.

Index	Binary	Reversed	B(Index)
0	000	000	0
1	001	100	4
2	010	010	2
3	011	110	6
4	100	001	1
5	101	101	5
6	110	011	3
7	111	111	7

done in-place to save storage. That is, the output replaces the input array in memory. The second step rearranges the positions output by the first step so that the correctly indexed frequency values are in their respective bins. In other words the first step produces the DFT but with scrambled indices in what is called "Bit Reversed Order". Let the output of step one be $\vec{X}_1$. Saying that $\vec{X}_1$ is in bit reversed order means that $\vec{X}_1(k)$ equals $\vec{X}[B(k)]$, where $B(\cdot)$ is the bit reversing operation and $\vec{X}$ is the DFT of the input $\vec{x}$.

The bit reversal is a natural consequence of the first stage of the FFT and will be commented on later. The best description of this operation is obtained by representing each index k in binary. The number of bits selected for this representation equals the number necessary to represent the value $N - 1$, where N is the length of the sequence. The binary representation is then turned around (reversed) so that the most significant bit becomes the least significant, etc. As an example consider a length eight sequence (N=8) which implies that the bit reversal requires three binary digits (bits) per index as given in Table C.1. Note that because $B[B(k)] = k$, the equality $\vec{X}_1(k) = \vec{X}[B(k)]$ implies $\vec{X}_1[B(k)] = \vec{X}(k)$. More important than the precise nature of this reordering is the relatively simple way it can be eliminated from FFT-based radar processing.

As was noted before, there exist several FFT and IFFT algorithms which perform the in-place DFT in step one and reorder the vector in step two. There are also FFT and IFFT techniques which reorder (bit reverse) the data in step one and perform an in-place DFT in step two. When these steps are separate computer routines, an FFT algorithm where the output is left in bit reversed order can be combined with an IFFT algorithm which accepts bit reversed data as input. This completely eliminates the bit reversals from the radar processing steps outlined at the beginning of the section. Using the superscript BR to denote sequences in bit reversed order, this new more efficient approach can be summarized as

1. Appropriately sample the reference $u(t)$ to obtain $\vec{u}$.

2. Take the FFT with BR output of $\vec{u}$ creating $\vec{U}^{BR}$.

3. Complex conjugate $\vec{U}^{BR}$ to create matched filter $\vec{U}^{*BR}$.

4. Take the FFT with BR output of $\vec{r}$ creating $\vec{R}^{BR}$.

5. Implement filter by multiplying $\vec{R}^{BR}$ with $\vec{U}^{*BR}$.

6. Convert to correctly ordered time sequence using IFFT with BR input.

It is important to realize that this adaptation is not always the most efficient approach to improving the performance of the FFTs. This is especially true when special purpose hardware or microcoded algorithms already exist and are available which would outperform the above adaptation independent of how efficient the software is written. Because this is a standard technique, however, many of these special FFT implementations include the bit reversal as a separate step which can be eliminated using this approach. One note of caution: if you decide to ambitiously create this environment on your own, then two different algorithms are necessary for the forward and the inverse FFTs to accommodate the use of bit reversed data as output and input.

C.2   FFT of a Real Valued Sequence

We begin this section by noting a powerful property of the discrete Fourier transform $\vec{X}$ of a complex valued sequence $\vec{x}$ of length $2N$.

$$
\begin{aligned}
\vec{X}(k) &= \sum_{n=0}^{2N-1} \vec{x}(n)\, e^{-j2\pi nk/(2N)} \\
&= \sum_{n=0}^{N-1} \vec{x}(2n)\, e^{-j2\pi(2n)k/(2N)} + \sum_{n=0}^{N-1} \vec{x}(2n+1)\, e^{-j2\pi(2n+1)k/(2N)} \\
&= \sum_{n=0}^{N-1} \vec{x}(2n)\, e^{-j2\pi 2nk/N} + e^{-j2\pi k/(2N)} \sum_{n=0}^{N-1} \vec{x}(2n+1)\, e^{-j2\pi nk/N} \\
&= \vec{X}_{even}(k) + e^{-j2\pi k/(2N)}\, \vec{X}_{odd}(k). \tag{C.1}
\end{aligned}
$$

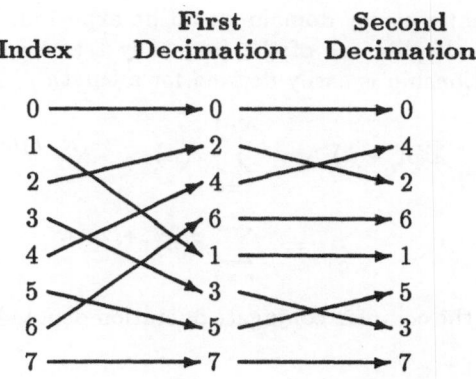

Figure C.1: Bit reversal by repeated decimation in time.

This implies that the length $2N$ discrete Fourier transform can be performed as two length N transforms—one performed on the even indexed points the other on the odd with an appropriate phase weighting. This reduction in sequence length is the basis for most FFT algorithms and is called decimation in time. The process is continued until sequences with lengths that allow efficient computation of the transform have been reached. For sequences with power of two lengths, the repeated decimation of time by even and odd selection of samples results in groupings based on the bit reversed pattern discussed previously and seen in Figure C.1. The dashed lines are used to denote the arrangement of the even positioned elements. The odd positioned elements move along solid arrows. For length eight sequences, the bit reversal is completed on the second time decimation. If the original sequence $\vec{x}$ had length pN, p some integer, then the DFT of $\vec{x}$ could be accomplished as the weighted sum of p transforms each of length N. This decimation procedure describes the internal structure of many FFT algorithms. For our purpose, the structure of an FFT is less important than a fast means of calculating the transform for a sequence which is real valued.

Clearly, one approach is to edit the existing FFT routine and remove all the multiplications and additions associated with the imaginary terms in the input since these are known to be zero. This is a relatively effective approach and will yield reasonable results. As was mentioned above, however, we are not interested in writing our own FFT algorithm if it is possible to utilize existing complex to complex FFT programs. Because all the imaginary terms in the input are zero, a reduction in the information contained in the frequency domain would be expected. This is in fact the case with one half plus one of the frequency terms specifying the entire array. The relationship is easily derived for a length N real input $\vec{x}$ as

$$
\begin{aligned}
\vec{X}(N - k) &= \sum_{n=0}^{N-1} \vec{x}(n)\, e^{-j2\pi(N-k)/N} \\
&= \sum_{n=0}^{N-1} \vec{x}(n)\, e^{+j2\pi k/N}
\end{aligned}
\tag{C.2}
$$

For real inputs, the complex conjugate operation denoted by * has no effect. This results in

$$
\vec{X}(N - k) = \left\{ \sum_{n=0}^{N-1} \vec{x}(n)\, e^{-j2\pi k/N} \right\}^{*} = \left\{ \vec{X}(k) \right\}^{*} = \vec{X}^{*}(k)
\tag{C.3}
$$

Consequently for N even, the first $N/2 + 1$ terms specify the entire sequence. Knowing that about half the terms are necessary to specify the entire DFT intuitively suggests that there might be some way to modify the input sequence to perform this operation using a length $N/2$ complex transform. We begin investigations into this by examining the decimation in time for a special complex input.

Consider the real sequence $\vec{x}$ of length $2N$. Let the length N complex sequence $\vec{y}$ be defined as

$$
\vec{y}(n) = \vec{x}(2n) + j\vec{x}(2n + 1), \quad \text{for } 0 \le n \le N - 1.
\tag{C.4}
$$

In many computer codes this is simply done by declaring $\vec{y}$ a complex array of length N, $\vec{x}$ a real array of length $2N$, and equivalencing the two arrays. Because complex arrays are usually stored sequentially as real and imaginary parts, this overlaying of the vectors performs the appropriate even-odd mapping to complex. The DFT of this complex sequence is denoted $\vec{Y}$ where

$$
\begin{aligned}
\vec{Y}(k) &= \sum_{n=0}^{N-1} [\vec{x}(2n) + j\vec{x}(2n + 1)]\, e^{-j2\pi nk/N} \\
&= \vec{X}_{even}(k) + j\vec{X}_{odd}(k).
\end{aligned}
\tag{C.5}
$$

Obviously $\vec{X}_{even}$ and $\vec{X}_{odd}$ are the N point transforms of the even and odd indexed points from the real vector $\vec{x}$, respectively. This implies that these

N point transforms have the symmetry property described in Equation C.3. This means that $\vec{Y}^*(N-k)$ equals $\vec{X}_{even}(k) - j\vec{X}_{odd}(k)$. Together with Equation C.5 this can be used to solve for $\vec{X}_{even}$ and $\vec{X}_{odd}$ in terms of the vector $\vec{Y}$. These relations are

$$\vec{X}_{even}(k) = \frac{1}{2}\left[\vec{Y}^*(N-k) + \vec{Y}(k)\right]$$
$$\vec{X}_{odd}(k) = \frac{1}{2}\left[\vec{Y}^*(N-k) - \vec{Y}(k)\right] \tag{C.6}$$

Combining these results for real sequences with the general result of Equation C.1, implies that

$$\vec{X}(k) = \vec{X}_{even}(k) + e^{-j2\pi k/(2N)}\,\vec{X}_{odd}(k). \tag{C.7}$$

Recall that the benefit of computing $\vec{X}_{even}$ and $\vec{X}_{odd}$ by Equation C.6 is a reduction, by a factor of two, in the required length of the complex FFT performed. Summarizing the steps for taking a DFT of a real valued sequence $\vec{x}$ of length $2N$ using complex to complex FFT's of length N:

1. Generate the complex sequence $\vec{y}$ by shifting the odd indexed elements of the real sequence $\vec{x}$ into the real parts of $\vec{y}$ and the even indexed elements into the imaginary parts. That is, $\vec{y}(n) = \vec{x}(2n) + j\vec{x}(2n+1)$, for $0 \leq n \leq N-1$.

2. Take complex to complex FFT of $\vec{y}$ creating $\vec{Y}$.

3. Compute $\vec{X}$, the DFT of $\vec{x}$ from $\vec{Y}$ using

$$2\vec{X}(k) = \vec{Y}^*(N-k) + \vec{Y}(k) + \left(\vec{Y}^*(N-k) - \vec{Y}(k)\right)e^{-j2\pi k/(2N)}$$

Again it is important to note that special hardware or system dependent FFT algorithms may outperform this technique. The extra control operations and development time must be considered when comparing the use of existing complex to complex algorithms with this adaptation.

C.3 Summary of Implementation Considerations

Each application, each set of radar data, and each computer will have peculiarities which can be exploited in order to improve the performance of the software algorithm. The time and expertise required to implement many of these modifications can be more costly than the savings provided by the increased speed of the system. For this reason, it is assumed in these notes that fast convolutions are performed with complex to complex FFT algorithms. It is recommended that the modifications required to implement the faster algorithms presented in this section should be attempted on a second implementation of a system. That is, after the initial system is debugged and providing valid output. Unless many data sets are to be processed, it may not be cost effective to pursue more sophisticated algorithms. In addition, extremely fast special purpose hardware for performing floating point digital signal processing operations (including FFTs) is available for applications which require high speed computation.

C.4   References

J. W.Cooley and J. W. Tukey, "An algorithm for the machine calculation of complex Fourier series," *Mathematics of Computation,* vol. 19, no. 90, pp. 297–301, 1965. Reprinted in *Digital signal processing,* ed. by L. R. Rabiner and C. M. Rader, IEEE Press, New York, 1972.

J. W. Cooley, P. A. W. Lewis, and P. D. Welch, "The fast Fourier transform algorithm: programming considerations in the calculation of sine, cosine and Laplace transforms," *J. Sound and Vibration,* vol. 12, pp. 315–337, July 1970. Reprinted in *Digital signal processing,* ed. by L. R. Rabiner and C. M. Rader, IEEE Press, New York, 1972.

C.5   Problems

PC.1 For a fixed signal length L, how long must the reference (filter coefficients) be before FFT processing is more efficient than direct implementation of the convolution?

Appendix D

Solutions to Exercises

1.1 Let $T = t_{down} + (T - t_{down})$ and let D be the depth of the well. Then
$$D = v_{rock}\, t_{down} = v_{sound}\, (T - t_{down}).$$

Solving for t_{down} and using in the equation for D results in
$$D = v_{rock}\, t_{down} = \frac{T\, v_{rock}\, v_{sound}}{v_{rock} + v_{sound}}$$

1.2 Using the same notation for time down:
$$\begin{aligned}
D &= \frac{1}{2}\, g\, t_{down}^2 = v_{sound}\, (T - t_{down}) \\
t_{down} &= \frac{1}{g}\left(-v_{sound} + \sqrt{v_{sound}^2 + 2gTv_{sound}}\right) \\
D &= \frac{1}{g}\, v_{sound}^2 + Tv_{sound} - \frac{1}{g}\, v_{sound}\sqrt{v_{sound}^2 + 2gTv_{sound}}
\end{aligned}$$

1.3 Coding the pulse (time-division or frequency-division multiplexing) can eliminate collisions if a each channel (time or frequency) only has one assigned transmitter. Collision detection and error correction can be incorporated by increasing the capacity (time or frequency) of the channel.

1.4 a) $A^*(-\tau) = \int u(t)\, u^*(t - \tau)\, dt = \int u(s + \tau)\, u^*(s)\, ds = A(\tau)$;
b) $\int A(\tau)e^{-j2\pi f\tau}\, d\tau = |U(f)|^2 \geq 0$, where $U(f) = \text{FT}[u(t)]$.   c) Expand $\int |u(t) + \epsilon u(t + \tau)|^2\, dt \geq 0$ assuming ϵ is a real number. Or notice that
$$|A(\tau)| = \left|\int |U(f)|^2 e^{j2\pi f\tau}\, df\right| \leq \int |U(f)|^2\, df = A(0)$$

1.5 For $\kappa = 2\pi a\tau$, the integral is

$$\int_{T_0}^{T_0+T-\tau} e^{j\kappa t}\, dt \;=\; \frac{1}{j\kappa}\, e^{j\kappa[T_0+.5(T-\tau)]}\left\{e^{.5j\kappa(T-\tau)} - e^{-.5j\kappa(T-\tau)}\right\}$$

$$\;=\; \frac{2}{\kappa}\, e^{j\kappa[T_0+.5(T-\tau)]} \sin[.5\kappa(T-\tau)],\; 0 \le \tau \le T$$

$$\int_{T_0-\tau}^{T_0+T} e^{j\kappa t}\, dt \;=\; \frac{1}{j\kappa}\, e^{j\kappa[T_0+.5(T-\tau)]}\left\{e^{.5j\kappa(T+\tau)} - e^{-.5j\kappa(T+\tau)}\right\}$$

$$\;=\; \frac{2}{\kappa}\, e^{j\kappa[T_0+.5(T-\tau)]} \sin[.5\kappa(T+\tau)],\; -T \le \tau \le 0.$$

1.6 For a pulse of duration T, centered about the origin:

$$A(\tau) = e^{j2\pi f\tau}\, \frac{\sin[\pi a\tau(T-|\tau|)]}{\pi a\tau}.$$

1.7 a) The autocorrelation of a rectangle function is a triangle function. This is easily shown with the following Fourier transform pairs.

$$r(t) = \begin{cases} A, & |t| \le T/2; \\ 0, & \text{else.} \end{cases} \qquad\qquad R(f) = AT\,\frac{\sin(\pi fT)}{\pi fT}$$

$$t(t) = \begin{cases} B(1 - |t|/T), & |t| \le T/2; \\ 0, & \text{else.} \end{cases} \qquad T(f) = BT\left(\frac{\sin(\pi fT)}{\pi fT}\right)^2$$

b) The autocorrelation is $\frac{1}{2a}\, e^{-a|t|}$ for a one-sided exponential function.

$$o(t) = \begin{cases} e^{-at}, & 0 \le t; \\ 0, & \text{else.} \end{cases} \qquad O(f) = \frac{1}{a+j2\pi f}$$

$$l(t) = e^{-b|t|} \qquad\qquad L(f) = \frac{2b}{b^2+(2\pi f)^2}$$

1.8 Following the same logic as for real-valued signals:

$$d^2 \;=\; \int [u^*(t) - v(t+\tau)]^2\, dt$$

$$\;=\; \int |u(t)|^2\, dt + \int |v(t+\tau)|^2\, dt$$

$$\;-\; \int u^*(t)\, v(t+\tau)\, dt - \int u(t)\, v^*(t+\tau)\, dt$$

$$2\mathrm{Real}[C(\tau)] \;=\; C(\tau) + C^*(\tau)$$

$$\;=\; \int u^*(t)\, v(t+\tau)\, dt + \int u(t)\, v^*(t+\tau)\, dt$$

1.9 Superposition and shift-invariance are shown by:

$$
\begin{aligned}
y_1(t) &= \int u^*(s)\, r_1(t+s)\, ds \\
y_2(t) &= \int u^*(s)\, r_2(t+s)\, ds \\
y_1(t) + y_2(t) &= \int u^*(s)\, r_1(t+s)\, ds + \int u^*(s)\, r_2(t+s)\, ds \\
&= \int u^*(s)\, [r_1(t+s) + r_2(t+s)]\, ds \\
y_1(t+\tau) &= \int u^*(s)\, r_1(t+\tau+s)\, ds
\end{aligned}
$$

1.10 Using Fourier transform properties:

$$
\begin{aligned}
a) \quad R(f) &= \Sigma(f)\, U(f) \\
\Sigma_{IF}(f) &= \frac{R(f)}{U(f)}, \ U(f) \neq 0 \\
\Sigma_{IF}(f) &= \frac{R(f)\, U^*(f)}{|U(f)|^2} \\
b) \quad \Sigma_{MF}(f) &= R(f)\, U^*(f) \\
c) \quad |U(f)| &\approx 1
\end{aligned}
$$

1.11 Want τ such that $\pm 1 = a\tau(T - |\tau|)$, (the first zero of the sine). For $\tau = 1/(aT)$ and $aT^2 \gg 1$,

$$
a\tau(t - |\tau|) = \frac{a}{aT}\left(T - \frac{1}{|aT|}\right) = 1 - \frac{1}{aT^2} \approx 1
$$

1.12 The Fourier transform of the analytic signal $u(t)$ is

$$
\begin{aligned}
U(f) &= U^r(f) + j\, U^h(f) \\
&= \begin{cases} U^r(f) - j^2\, U^r(f), & \text{if } f > 0; \\ U^r(f) + 0, & \text{if } f = 0; \\ U^r(f) + j^2\, U^r(f), & \text{if } f < 0. \end{cases} \\
&= \begin{cases} 2U^r(f), & \text{if } f > 0; \\ U^r(f), & \text{if } f = 0; \\ 0, & \text{if } f < 0. \end{cases}
\end{aligned}
$$

Because our example has $U^r(0) = 0$, the FFT peforms these operations by setting the negative frequencies to zero (half the points) and ignoring the scaling by the constant 2. Note that multiplying by j is equivalent to 90 degree shifts.

2.1 $L = \sqrt{8R \cdot \delta r + 4(\delta r)^2}$

2.2 $\theta = 2\pi(2\delta r)/\lambda$ or $\delta r = \theta\lambda/(4\pi)$.

2.3 $L_{unfocused} = \sqrt{8R \cdot \delta r} = \sqrt{R\lambda/8}$ which implies that the resolution is $\rho = \lambda R/(2L) = \sqrt{2\lambda R}$.

2.4 $\delta r = \sqrt{(L/2)^2 + R^2} - R = R\left\{\sqrt{\left(\frac{\lambda}{2D}\right)^2 + 1} - 1\right\}$

2.5 Let $R = 850$km, $\lambda = 23.5$cm, and $D = 10.5$ or 2m, then

$$L_{focused} = \frac{\lambda R}{D} = \begin{cases} 100\text{km}, & \text{in range (2m orientation);} \\ 19\text{km}, & \text{in azimuth (10.5m orientation).} \end{cases}$$

The actual measured 3dB footprint is 100km by 15km which implies a range variation and phase delay of

$$\delta r = \sqrt{\left(\frac{L}{2}\right)^2 + R^2} - R = 33.1\text{m}$$
$$\phi = 2\pi(2\delta r)/\lambda = 2\pi \cdot 281.5\,\text{radians}$$

2.6 $f_{D_0} < 0.0$ implies a negative radial velocity.

n_shift = abs(f_{prf} * sampling_interval/(f_{D_0} * λ_c * n_references))
if (n_count $\geq$ n_shift) then
    n_ref = n_ref - 1
    if (n_ref < 1) then
        n_ref = n_references
        i_start = i_start - 1
    endif
endif

2.7 Determine the starting frequency for a look.

```
          t_SA = len_az_ref * n_looks / f_prf
c         determine center frequency of each look
          f = f_D + (2.5 - look) *ḟ_D* (t_SA / n_looks)
          f = modulo( f, f_prf )
          if ( f < 0.0 ) f = f + f_prf
          if ( f > (f_prf/2) ) f = f - f_prf
          i_az_freq = len_az_array + f * len_az_array / f_prf
c         convert center index to starting index for look
          i_az_freq = modulo( i_az_freq-(len_look/2), len_az_array )
          if ( i_az_freq < 1 ) i_az_freq = i_az_freq + len_az_array
```

3.1 At f, see Equation 3.1.

3.2 $C(u,v) = A(u)B(v)$, why isn't this a convolution?

3.3 a) $S(u,v)\,e^{-j2\pi(x_0u+y_0v)}$ is the FT of $s(x-x_0,y-y_0)$.   b) $S(u/a,v/b)/|ab|$ is the FT of $s(ax,by)$.   c) $S(u\cos\theta+v\sin\theta,-u\sin\theta+v\cos\theta)$ is the FT of $S(x\cos\theta+y\sin\theta,-x\sin\theta+y\cos\theta)$—rotation by θ in space produces equal rotation in frequency. For fun try $f(ax+by,cx+dy)$.

3.4 $S(u,v)=S^*(-u,-v)$ which implies the real part of S is even, the imaginary part of S is odd, and $|S|$ is even.

3.5 Converts a real-valued image (transparency) to a complex-valued one.

3.6 Horizontal stripes appear along the vertical axis of the frequency domain (why?). An appropriate filter blocks these positions—a pair of needles taped to a transparency works well in an optical system.

3.7 Dark ground is short for dark background, with no input transparency to modulate the input plane wave there is no light out—the plane wave is focused to DC in the frequency domain.

4.1 Use the transducer pattern as a reference.

4.2 By the projection-slice theorem all slices intersect at the origin (DC) in two-dimensional frequency space. Implies that the average transmission is the same independent of viewing angle.

4.3 Note $S(\rho,\theta+\pi)=S(-\rho,\theta)$ in polar format representation, then invoke the projection-slice theorem:

$$\sigma(x,y) = \int_0^{2\pi}\int_0^\infty S(\rho,\theta)e^{j2\pi\rho(x\cos\theta+y\sin\theta)}\rho\,d\rho\,d\theta$$

$$= \int_0^\pi\int_{-\infty}^\infty S(\rho,\theta)e^{j2\pi\rho(x\cos\theta+y\sin\theta)}|\rho|\,d\rho\,d\theta$$

4.4 The projections $p_\theta(t)$ contain the same information in 0 to π as they do in π to 2π. The line integrals are the same because the part is just turned around 180 degrees.

4.5 Using the residue theorem from complex variables,

$$FT(|\pi r|^{-1}) = \int_{-\infty}^\infty |\pi r|^{-1}\,e^{j2\pi r\rho}\,dr$$

$$= \pi(\rho\lim_{r\to0}\pi^{-1}e^{j2\pi r\rho})$$

4.6 Let $H_L(f)$ be one for $-L \leq f \leq L$ and zero elsewhere. Then

$$
\begin{aligned}
h(t) &= \int_{-L}^{L} |f| e^{j2\pi ft} \, df \\
&= 2 \left[\frac{\cos(2\pi ft)}{(2\pi t)^2} + \frac{f \sin(2\pi ft)}{2\pi t} \right]_0^L \\
&= 2 \frac{\cos(2\pi Lt) - 1}{(2\pi t)^2} + +2 \frac{L \sin(2\pi Lt)}{2\pi t} \\
&= -L^2 \left[\frac{\sin(\pi Lt)}{\pi Lt} \right]^2 + 2L^2 \frac{\sin(2\pi Lt)}{2\pi Lt}
\end{aligned}
$$

4.7 $\sigma_1 = 4, \sigma_2 = 9, \sigma_3 = 6$, and $\sigma_4 = 1$.

4.8 The reference for mixing is $2\cos\{2\pi[f_c(t - \tau_0) + .5a(t - \tau_0)^2]\}$ and the product with $r(t)$ is

$$
\int_{-w}^{w} p_\theta(\tau) \left[e^{j2\pi\{f_c(2t - \tau - 2\tau_0) + .5a(t - \tau_0)^2 + .5a(t - \tau - \tau_0)^2\}} \right.
$$
$$
\left. + e^{j2\pi\{.5a\tau^2 - [f_c + a(t - \tau_0)]\tau\}} \right] d\tau.
$$

A.1 $X(0) = 2$, $X(1) = 1 - j$, $X(2) = 0$, and $X(3) = 1 + j$.

A.2 Note that $Z(0) = 4$, $Z(1) = 2$, $Z(2) = 0$, and $Z(3) = 2$. Therefore, $z(0) = 1$, $z(1) = 2$, $z(2) = 1$, and $z(3) = 0$.

A.3 Outputs are the same.

A.4 Use the same procedure as in the previous problems.

B.1 Note that $R_{NN}(\sigma, \tau) = R_{NN}(\tau, \sigma)$.

B.2 $H(\omega) = F\{h(t)\} = S(-\omega)e^{-j\omega T}$.

B.3 The convolution integral for the matched filter can be put in the form of a correlation integral.

$$
\begin{aligned}
x(t) * h(t) &= \int_{-\infty}^{\infty} x(\tau) h(t - \tau) \, d\tau \\
&= \int_{-\infty}^{\infty} x(\tau) s(T - (t - \tau)) \, d\tau \\
&= \int_{-\infty}^{\infty} x(\tau) s(\tau + (T - t)) \, d\tau
\end{aligned}
$$

B.4 The complex conjugate of the time reversed signal.

B.5 If $N(t)$ is wss then by definition

$$E\{|N(t)|^2\} < \infty;$$
$$R(t,s) = E\{N(t)\,N(s)\} = R(t-s)$$

for all t and s. The matched filter equation for a non-causal filter becomes

$$\int_{-\infty}^{\infty} R(\tau - \sigma)h(T - \sigma) = s(\tau), \quad -\infty < \tau < \infty.$$

Fourier transforming both sides results in

$$H(\omega) = \frac{S(-\omega)}{\phi_N(\omega)}e^{-j\omega T}$$

where ϕ_N is the power spectral density associated with R_N.

C.1 When L is much greater than N, the convolution is best performed by using FFTs on sections of the data (each of length near N) and then combining the short section outputs to get the composite larger convolution output.

Index

A-scan, 34
absorption function, 114
Acousto-Optic cell
    A-O, 106
Algebraic Reconstruction Technique
    ART, 117, 120
aliasing, 132
along track, 35
antenna array
    real focused, 38
    real unfocused, 37
    real, 37
    synthetic, 39
apodization, 92
attenuation function, 114
autocorrelation function, 5
azimuth, 35
azimuthal phase factor, 52
B-scan, 34
backprojection, 118
base band, 12
beating, 13
bit reversed order, 155
carrier frequency, 11
Charge Coupled Device
    CCD, 106
chirp, 6
    rate, 6
coherent, 13, 90
compression ratio, 8
convolution, 137
    circular, 140, 143
    periodic, 140, 143

sum, 138
cut-and-paste, 67
decimation in time, 157
Discrete Fourier Series
    DFS, 139
Discrete Fourier Transform
    DFT, 142, 22
Doppler shifts, 57
Doppler
    frequency, 56
    rate, 56
Fast Fourier Transform
    FFT, 139, 142, 23
filters
    high pass, 12, 91
    knife edge, 91
    low pass, 91
    matched, 10, 147, 150
footprint, 35
foreshortening, 112
Fourier transform, 15
Fresnel lens, 99
ground range, 35
Hilbert transform, 15
hologram, 104
homogeneous, 110
I-Q, 15
imaging equation, 86
in-phase, 15
isotropic, 110
layover, 112
lenses
    conical, 95

169